Dynamics of Agricultural Information Management

(Prof. P. A. Shinde Festschrift)

The Editors

Dr. G. Rathinasabapathy is serving as the University Librarian of Tamil Nadu Veterinary and Animal Sciences University (TANUVAS), Chennai. He was awarded M.Com., M.L.I.S., by Annamalai University, M.Phil. (LIS) by Alagappa University and Ph.D. (LIS) by the University of Madras. He has 23 years of professional experience and contributed about 110 research papers to journals, International and National Seminar volumes and participated about 35 International/National level programmes in LIS. He has authored/edited four books and five volumes. He has obtained three projects worth about Rs.300 lakhs for the development of Library and Information facilities at TANUVAS. He is a member in the Editorial Board of Seven Peer Reviewed Journals and presently the Editor-in-Chief of the Indian Journal of Agricultural Library and Information Services (IJALIS) of AALDI. He is a reviewer for many International Journals including 'Scientometrics' and "PLOS" journal. He is the elected President of Tamil Nadu Library Association, Secretary of MALIBNET and life member of 18 professional societies viz., ILA, IASLIC, MALA, AALDI, SALIS, SLA (USA), ISTE, CSI, etc. He has served as the Chairman and Member of about 65 Committees at National and Regional level. Delivered about 20 invited talks and organized many national level Continuing Educational Programmes. He visited many leading libraries in the USA. He was honoured with many awards including 'Award for Best Paper Presentation' by the Indian Society for Veterinary Pharmacology and Toxicology (2005), "Useful Writer Award" (2012) by Tamil Nadu Magazine Publishers Association, 'Bharat Gaurav Award (2016)' by an International Social Service Association, New Delhi and "Certificate of Appreciation" by ICAR for his contribution to 'KrishiKosh" Institutional Repository(2017).

Dr. K. Veeranjaneyulu, M.Com.,M.L.I.S.,BGL, PGDLAN, Ph.D., is currently the University Librarian & Professor of Prof.Jayashankar Telangana State Agricultural University, Hyderabad. He has more than two and half decades of professional experience in the field of library and information service and a Resource person to various Academic Staff Colleges in the country, and delivered many Guest Lectures in Universities of Andhra Pradesh, Maharashtra, Tamil Nadu, Karnataka and Haryana. He has participated in more than 100 National Conferences, Workshops and Seminars, and written and edited more than 25 books and contributed around 150 articles. He is a member of Boards of Studies, Academic Councils and Faculty Boards of many Universities in India, and a Ph.D. Adjudicator and External Examiner for several Universities. Under his guidance, ten candidates were awarded M. Phil. Degrees and two Ph.D. Currently he is guiding six Ph. D. students. He is a life member of various learned societies, and Chief Editor of the "Indian Journal of Agricultural Librarians & Information Services" (IJALIS). He visited many foreign countries including the USA as part of his academic and research assignments. He has been awarded USSHLE-IJILS Param Bhusan B.B.Shukla Millennium Award for excellence in 2007. He received "Certificate of Appreciation" awarded for outstanding contribution in developing KrishiKosh (Agricultural Institutional Repositories Database) in the Libraries of NARES under the e-Granth Project from the Director General, ICAR and National Director, NAIP, IARI, New Delhi. He also received Appreciation Certificate from ICAR for uploading high number of theses i.e. more than 5000 in the Krishikosh during the year 2016.

Dr. Kishor N. Patil, M.A., MLIS, Ph.D., is serving as the Librarian, College of Agriculture, Dhule under Mahatma Phule Krishi Vidyapeeth, Rahuri. He has 31years of professional experience to his credit. He has published many research papers in journals and presented papers in conferences and contributed many book chapters. He has attended many National and International conferences/Workshops /Seminars. Presently, he is Life member of AALDI, PULISA, ILA, and IASLIC. He has organized many User Orientation/Awareness Programmes in the University for the benefit of library users.

Dynamics of Agricultural Information Management

(Prof. P. A. Shinde Festschrift)

— Editors —

G. Rathinasabapathy
K. Veeranjaneyulu
Kishore N. Patil

2018

Daya Publishing House®

A Division of

Astral International Pvt. Ltd.

New Delhi – 110 002

ISBN 9789387057340 (Int Edition)

Published by : **Daya Publishing House®**
A Division of
Astral International Pvt. Ltd.
– ISO 9001:2008 Certified Company –
4736/23, Ansari Road, Darya Ganj
New Delhi-110 002
Ph. 011-43549197, 23278134
E-mail: info@astralint.com
Website: www.astralint.com

Digitally Printed at : **Replika Press Pvt. Ltd.**

Mahatma Phule Krishi Vidyapeeth

Rahuri, Dist. Ahamadnagar - PIN- 413 722, Maharashtra State, INDIA

Foreword

Dr. K.P. Viswanatha
Vice-chancellor

The Mahatma Phule Krishi Vidyapeeth, Library is one of the premier Library around the State Agricultural Universities across the country. Since its inception, it has been catering to the needs of students, faculty members and the various agricultural entrepreneurs.

University Library has undergone a demand driven change and has been bestowed upon a certificate of appreciation by the Indian Council of Agricultural Research, New Delhi for its valuable contribution to KrishiKosh. It is a versatile open access digital repository catering to the needs of NARES. The University Librarian has also been awarded the certificate of appreciation by the ICAR, New Delhi for his commendable contribution for strengthening and sustainability of e-Granth.

It gives me immense pleasure in bringing out a Festschrift volume in honour of the University Librarian, Prof. P.A.Shinde who is retiring from the university service after putting on 36 years of yeomen service in the field of library and information services.

With best wishes,

(K. P. Viswanatha)

Preface

We are immensely pleased to dedicate this Festschrift to Prof. Prakash Arjunrao Shinde, the University Librarian of Mahatma Phule Krishi Vidyapeeth (MPKV), Rahuri on the eve of his superannuation.

Prof.P.A.Shinde was born on 21st July 1955. After finishing his B.A. and B.L.I.S. in Marathwada University, Aurangabad he started his career in 1982 at College of Agriculture, Pune and rose to the position of University Librarian of Mahatma Phule Krishi Vidyapeeth, Rahuri which is known as one among the leading agricultural library of India.

The contributions made by Prof.P.A.Shinde to the development of library and information services at MPKV are laudable as with very limited technical human resource, he has done a commendable job to transform the book-only library into a modern digital library with the state-of-the-art technologies.

He has better utilized the 'Strengthening the Digital Library and Information Management under NARS (e-Granth)" scheme awarded under National Agricultural Innovation Project (NAIP) to convert the library into a modern digital library by using cutting edge technologies. His painstaking efforts to digitize the rare collections of the library are really a great contribution to built the national level digital repository.

When we have planned to honour Prof.P.A.Shinde with a Festschrift and discussed about this with the fellow agricultural librarians during an AALDI meeting, there was overwhelming response from them and many of them opined that this has to be done to honour Prof.P.A.Shinde as he is a gem of person among us.

Besides his professional contribution, he is a man with many good qualities who always very kind to others. He is very simple, humble, pious and always respects the values. Even during tough times, he never hurts anybody and always want to settle issues amicably. It is very rare to see a person like P.A.Shinde. We are sure that the professional contributions made by Prof.P.A.Shide will be remembered ever by the stakeholders of agricultural librarianship in India.

With profound respects and regards, we honour Prof.P.A.Shinde with this Festschrift which contains 25 articles dealing with various aspects of dynamics of agricultural information management. We are very happy to note that there is national level participation in this venture. We are very sure that the topics covered are also varied and very useful to the practicing librarians.

Dr. G. Rathinasabapathy
Dr. K. Veeranjaneyulu
Dr. Kishore N. Patil

Contents

REMINISCENCES OF PROF. P.A. SHINDE

Prof. P.A. SHIDE: A GEM OF PERSON

It gives me pleasure to draw a brief pen portrait of my friend, Prof. Prakash Arjunrao Shinde, the University Librarian of Mahatma Phule Krishi Vidyapeeth (MPKV), Rahuri. I came in contact with him in early 2009 when I was drafting the consortium design of the NAIP project which we later named it as e-granth. The name is a hybrid of modern with tradition. Granth is our sacred scriptures. The prefix e- denotes modernization of our learning resources. Astu.

The College of Agriculture, Pune was celebrating its centenary in 2008. In honouring this great achievement, the Government of India sanctioned a centenary special grant of Rs 100 crores. The ICAR desired to strengthen the MPKV university library through this e-granth consortium. Only a dozen libraries were chosen initially as partners in this consortium. This library was one among them and only the university from Maharashtra. As Consortium Principal Investigator, I was keen that or partners respond to our call for cooperation and commitment in implementation. During this process of discovery, I found Prof. Shinde to be a gem of person. He is a thorough professional in library management. My initial apprehension evaporated soon after interacting with him.

Even though soon after grand launch of this consortium, I moved to the Headquarters as the ADG in-charge of Education Planning and Development, I continued to be in touch with him in strengthening the libraries of both the university and its constituent colleges with funding from the Education Division. I also discovered while being a consultant of the NAIP that he contributed tremendously to the success of CeRA and other library projects.

Thanks to the personal efforts of Prof. Shinde, the MPKV library stands tall among the agricultural libraries. He is a team person and his contribution in strengthening AALDI vouchsafes my assertion. I am sure, his services and wise counsel will be utilized by MPKV and other knowledge generating systems. He is superannuating from his chequered innings of service.

I wish him all the best in terms of health, happiness and success in his lifelong service in the field of knowledge management.

Dr. C. Devakumar, FNAAS
Former FAO TCDC Consultant (Quality Assyrance)
Former Assistant Director General, ICAR,
New Delhi - 110012

Prof. P.A. SHINDE: AN EXCELLENT WARM HUMANBEING

I got an opportunity to work with Shri P.A. Shinde, University Librarian, MPKV, Rahuri during a world bank funded project under National Agricultural Innovation Project (NAIP), Strengthening of NARS libraries and digital knowledge management (eGranth). I found in him a highly dedicated and professionally committed Librarian, concerned for providing best possible library services to students and faculty. He rightly beleived that good library services can enhance the standards of teaching and research in the university.

Working on eGranth project, he contributed in standardization of library catalog and making it online accessible by creating OPAC and also sharing with other universities in NARS through AgriCat - an online union catalog of NARS. He and his team made huge contribution towards digitization of thesis, old valuable books, journals, university publications, important lectures and other useful content. This digitized content is made online accessible through a Digital Repository 'KrishiKosh'.

I was impressed to visit his library which was very well maintained with focus on students. A 24 hr reading room facility open to student truly made difference in student's academic life and career development. He organised several trainings, workshops and interactions to popularise digital library and access to digital resources for benefit of his students and faculty. He also participated along with his library team in several advance capacity building workshops and conferences to enhance the capacity of library to manage advance digital tools and resources for benefit of his library's stake holders.

During the six year period of active interaction with Shri P.A. Shinde through eGranth project, I always found him an excellent warm human being. I really enjoyed working with him and wish him happy, healthy and active life after superannuation. I am sure that professionally committed person like Shri Shinde will not sit idle and shall continue to professionally contribute towards advancement of role of library in present day technology oriented world.

Dr. A.K. Jain
Former Head, AKMU
India Agricultural Research Institute,
New Delhi - 110012

Prof. P.A. SHINDE: A LIBRARIAN EXTRAORDINARY

"Real leaders are ordinary people with extraordinary determination"

It is my pleasure in recollecting my memories with Prof. Prakash A. Shinde of Mahatma Phule Krishi Vidyapeeth, Rahuri and feel great privilege extended to me for expressing my opinions. Prof. Shinde is a friend with commitment, professional ethics, helping wherever it is needed.

Prof. Shinde is reflection of Dr. S.R. Ranganathan's Principle of "Hospitality in array and chain". He is quite hospitable to any one and never said anything negative. Professionals and friends across the country, who are visiting the "Holy place Shiridi" experience the hospitality shown by him.

Prof. Shinde is well known for his friendly attitude, library management skills, helping nature and intellectual thinking. He treated his seniors with deep regards and always received high regards from colleagues and students. Because of his good qualities such as simplicity, honesty, devotion, dedication towards profession I always treat him as an Extraordinary Librarian.

Prof. Shinde is a multi-faced personality who will always be remembered by his friendly behaviour. He is a burning example of "simple living and high thinking". Because of his friendly co-operation and helping nature, his circle of friends and fans are very large. His resourcefulness, enthusiasm, creativity, sincerity and foresightedness were extraordinary.

Dr. K. Veeranjaneyulu
University Librarian & Professor
PJTS Agricultural University
Hyderabad – 500 030, Telangana State

1

Information Retrieval (IR) through Text / Data Mining (TDM)

Prof. B Ramesh Babu

(Professor (Retd), University of Madras, Chennai &
Former Visiting Professor, Mahasarakham University, Thailand)
22/20B, Thangavelu Pillai Garden, First Street
Old Washermen Pet, Chennai 600 021
Email: beerakarameshbabu@gmail.com

Dr. C.K. Chandrasekhar

New No. 42/5, "DWARAKA"
Kamdarnagar, First Main Road
Nungambakkam, Chennai 600 034
Email: choudur@gmail.com

1. Introduction

Internet has brought in a revolution in the way we create, store, access, share and use information. Web and search engine technology are placing abundance of knowledge at our finger tips and yet we are unable to use all the information due to its shear volumes and are faced with the daunting task of going through the maze of useless information! The need to organize the huge amount of information led to the development of various organizational tools like classification schemes, catalogues, indexes, etc. The idea behind developing these organizational tools was to help users to have easy access and quick retrieval of relevant documents in the library. To find relevant information within this huge amount of information, there is an urgent need to organize these resources.

One of the components of Content Management System (CMS) is Indexing, search, and retrieval. For data to be valuable, it must be relevant to the task at hand and accessible in a timely fashion. Documents can be parsed for keywords, headings, graphics, and other elements; mechanisms for processing search requests become critical. More generally, effective content management systems support an organization's business processes for acquiring, filtering, organizing, and controlling access to information. The features of a CMS system vary, but most include Web based publishing, format management, revision control, and indexing, search, and retrieval (Ramesh Babu, 2011). This calls for text mining or data mining or Text/ Data mining (TDM). It has become an important research area, which refers to the application of machine learning (or data mining) techniques in the study of Information Retrieval and Natural Language Processing.

2. Information Retrieval

Information Retrieval (IR) is an interactive process. IR technologies are believed to be powerful tools to bridge the digital divide, as they can allow communities in developing countries to have access to timely and relevant information. However, this can only be realized if we understand the current information access practices of these communities. Then only we can design information access technologies that will enable these communities to bridge the digital divide. IR is a topic more associated with online documents and search engines. On the net there is a vast collection of documents. Only one or a few of them are of interest to any one individual. The goal of IR is to retrieve those documents that satisfy the user's need. So the input to the IR system is the clues that specify the user need. These clues are called keywords or a query. The basic concept of IR system is measuring similarity. A comparison is made as to how similar two documents are. Similar to text categorization method, the IR process transforms the user query into a document, matches the query document to each of the documents in the collection and retrieves the matched documents from the collection. IR is finding a document of an unstructured nature usually text that satisfies an information need from within large collections usually stored on computers. It is fast becoming the dominant form of information access, overtaking traditional database style searching and can also cover other kinds of data and information problems.

An information retrieval system deals with various sources of information on the one hand and users' requirements on the other. It must:

- Analyse the contents of the sources of information as well the users' queries, and then
- Match these to retrieve those items that are relevant.

These major functions of an information retrieval system can be listed as follows:

1. To identify the information (source) relevant to the areas of interest of the target users' community.
2. To analyse the contents of the sources (documents)
3. To represent the contents of the analyzed sources in a way that will be suitable for matching users' queries.
4. To analyze users' queries and to represent them in a form that will be suitable for matching with the database.
5. To match the search statement with the stored database
6. To retrieve the information that is relevant, and
7. To make necessary adjustments in the system based on feedback from the users (Chowdhury, 1999).

A diagrammatic representation of the Information retrieval (IR) process (Fig. 1) would appear as:

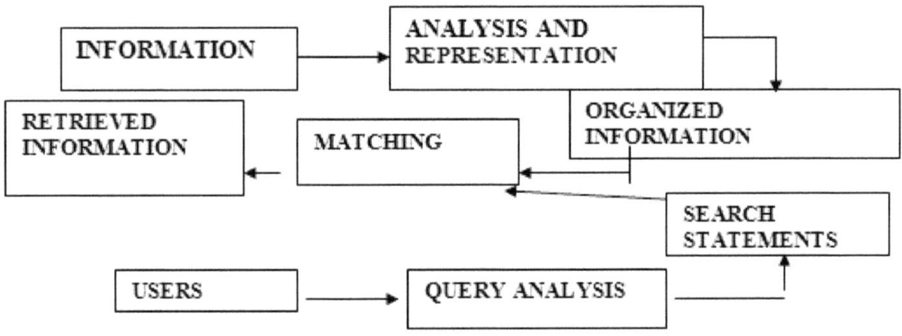

Fig.1 The IR Process

(Source: Chowdhury, G.G. Introduction to modern information retrieval. London: Library Association, 1999,p.4)

3. Concept of Text /Data Mining (TDM)

The term *"mining"* came to be associated with analysis of textual data because such data tends to be large. Processing a document running into several pages and containing thousands of words and other characters, numbers, addresses, and punctuations is inherently a complex task. Since it is difficult to cull out semantic information from documents except in some very specific domains, most text mining applications focus on word associations in documents i.e., use of co-occurrence of terms amongst documents as a measure of similarity and on that basis segregate the collection into groups of similar documents. Text mining and text analytics are broad umbrella terms describing a range of technologies for analyzing and processing semi- structured and unstructured text data. Text mining, also referred to as text data mining, roughly equivalent to text analytics, is the process of deriving high-quality information from text. It is the discovery of new previously unknown information from different written sources by automatically extracting that information by means of computer.

The phrase "text mining" is generally used to denote any system that analyzes large quantities of natural language text and detects lexical or linguistic usage patterns in an attempt to extract probably useful (although only probably correct) information (Sebastiani, 2002). It is the process of discovering useful and interesting knowledge from unstructured text. Text Mining, also known as Knowledge Text Analysis, Text/ Data Mining (TDM) or Knowledge-Discovery in Text (KDT) refers generally to the process of extracting information and knowledge from unstructured text. High-quality information is typically derived through the devising of patterns and trends through means such as statistical pattern learning. It is the process of deriving high quality information from data in the form of text written in natural languages. It is also referred to as text engineering, text data mining or text analytics (Grimes, 2007). It invokes statistical algorithms, natural language technologies, high speed computing and ability to handle large amounts of data in successfully developing and applying mathematical methods to unstructured text (Alpaydin, 2004). It endeavors to automate the process of reading textual content in documents leading to segmentation, information extraction and other types of analysis (Chandrasekhar, 2012). It belongs to the realm of big data where entire data from the population is used for analytical / inferential purposes. TDM strives to bring it out of the text in a form that is suitable for consumption by computers directly, with no need for a human intermediary.

Extraction of knowledge from unstructured written text is referred to as TDM and holds an enormous promise. A key element of text mining process is to link together information from various sources to form new information to be explored further by more conventional means of information processing. This is different from a web search. Typically in a web search the user is looking for something already known and written down. This something is hidden in a vast ocean of "www" data and the goal is sieve away unwanted information to extract the required information. This form of information extraction is called Information Retrieval (IR) and the tool used for this purpose is called the Search Engine. The key steps involved in information retrieval are shown in Fig. 2.

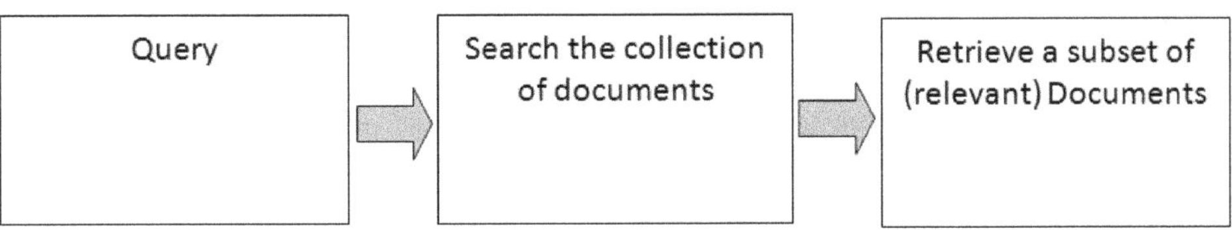

Fig: 2. Steps in Information Retrieval

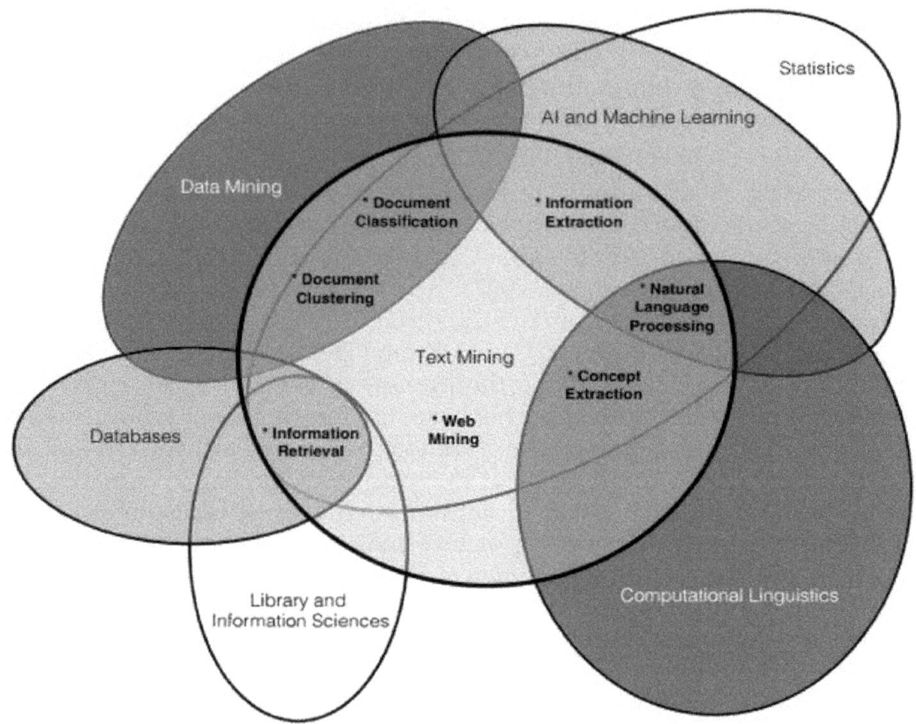

Fig. 3 Venn diagram for TDM

Source: https://www.pinterest.com/pin/541980136378711566/

Text and data mining (TDM) is the process of deriving information from machine-read material. It works by copying large quantities of material, extracting the data, and recombining it to identify patterns. Text and data mining (TDM) includes various technologies for the computer-based organisation and analysis of text and data. The goal of TDM is to discover new knowledge from already existing knowledge. It also helps to sort the vast amounts of information that organisations today rely on. 21st century data driven innovation and research relies on computers being able to analyse the vast amounts of information available digitally. In the internet environment, characterised by an abundance of information in a diversity of forms, text and data mining has become an essential tool for researchers and innovators (IFLA Statement on Text and Data Mining, 2013).

There are four stages to the TDM process (Fig.4). First, potentially relevant documents are identified. These documents are then turned into a machine-readable format so that structured data can be extracted. The useful information is extracted (Stage 3) and then mined (Stage 4) to discover new knowledge, test hypotheses, and identify new relationships.

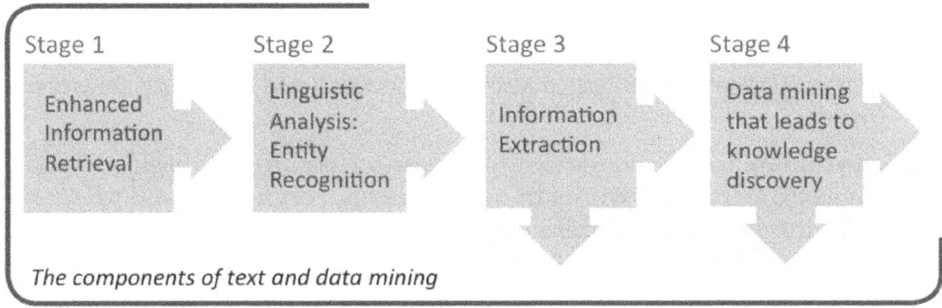

Fig. 4 Components of TDM

Source: http://libereurope.eu/text-data-mining/

Clark (2013) notes four main reasons to engage in TDM:

1. *To enrich content* -- Mining can improve indexing, be deployed to create relevant links, to improve the reading experience.
2. *Systematic review of literature* -- Mining can help researchers systematically review larger bodies of content, faster than they could do it themselves and to keep up with their field, without missing relevant information.
3. *Discovery* -- Mining can be used to create databases that can themselves be mined.
4. *Computational Linguistics research* -- Mining itself is the subject of research, for example to improve the extraction of meaning from texts.

4. Text /Data Mining Tasks

TDM is the process of seeking or extracting the useful information from the textual data. It tries to find interesting patterns from large databases. It uses different pre-processing techniques likes stop words elimination and stemming. The most important task in text/data mining applications is finding structure in a document collection (Chandrasekhar and Ramesh Babu, 2012).

In TDM, the goal is to discover heretofore unknown information, something that no one yet knows and so could not have yet written down.

- Document Search and Retrieval
- Document Clustering and Text Categorization
- Text Summarization
- Feature Identification and Extraction
- Authorship Ascription
- Language Identification
- Identification of Document Structure, Keywords, Key Phrases
- Identification of other entities like keywords, names and abbreviation

- Acronym Location and Definition
- Extraction of information for Template Fillings
- Learning template filling, document matching and similarity measurement rules.

Text /Data mining tools have started to appear for performing basic text analysis tasks such as extracting keywords from documents, condensing and presenting a summary in the form of a synopsis or abstract. It is embracing a lot of statistical methods in trying to assign semantics or meaning to parts of speech (Chandrasekhar and Ramesh Babu, 2012).

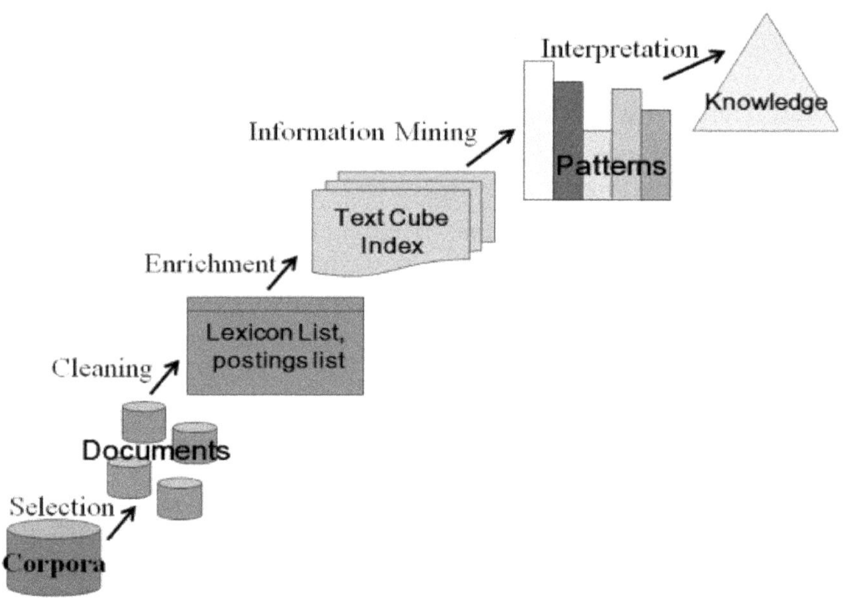

Fig. No. 5 Text /Data Mining Process

Source: http://article.sapub.org/image/10.5923.j.ac.20110101.01_001.gif

4.1 Techniques of Text /Data Mining

TDM appears to embrace the whole of automatic natural language processing and, arguably, far more besides—for example, analysis of linkage structures such as citations in the academic literature and hyperlinks in the Web literature, both useful sources of information that lie outside the traditional domain of natural language processing. But, in fact, most text/data mining efforts consciously shun the deeper, cognitive, aspects of classic natural language processing in favor of shallower techniques more akin to those used in practical information retrieval (Witten, 2005). Information retrieval might be regarded as an extension to document retrieval where the documents that are returned are processed to condense or extract the particular information sought by the user. Thus document retrieval could be followed by a text summarization stage that focuses on the query posed by the user, or an information extraction stage using techniques. Text/Data mining is not a single technique but a suite of techniques. It is a technique which extracts information from both structured and unstructured data and also finding patterns. TDM techniques are used in various types of research domains like natural language processing, information retrieval, text classification and text clustering.

The following techniques are generally considered as a part of Text /Data mining suite.

4.1.1 Automatic Cluster Detection:

Text classification is the process of organizing documents in pre-defined folders such that each folder contains similar documents. The folders might have been appropriately labeled and documents organized into them by a domain expert who knew how to classify the corpus of documents. There is always a possibility that the folders were not pre-defined and the required domain expertise is not available or accessible. Clustering is an automated technique that does not require a human expert to pre-create the classes nor to program rules for classification. For this reason the technique is called unsupervised machine learning technique. Clustering uses similarity / dissimilarity between documents to classify them into folders. Clustering techniques may be used in digital library to classify documents or to retrieve documents matching the profile constructed from user requirements.

4.1.2 Prediction and Trend Analysis:

Using this technique documents can be linked and sequenced on a time scale automatically using document creation time, citation and cross referencing information. When this is done we can analyze the trend and direction of the progress of theme of the subject matter and in a way predict the direction a new idea or concept is taking. This can be an excellent value-add for digital libraries in presenting documents in a historical sequence and accessing the rate of idea development.

4.1.3 Association Rules Mining:

Association analysis estimates co-occurrence of key terms in documents using measures such as support, confidence and lift. This technique establishes causal or temporal association between documents. This feature can be extremely useful in providing end users with suggestions on related documents or for paving a chronological road map.

4.1.4 Concept Description:

Concept description is to describe and summarize the relevant characteristics of document information objects through filtering, analyzing and comparing. Concept description is divided into similarity description and discrimination description. The former describes the object features commonality, while the latter describes the differences between the objects.

4.1.5 Temporal analysis and forecasting

This technique is similar to the regression technique with the independent variable being time. It models a relationship between independent and response variables using a training data set at known time periods. Once the model is built it may be used to estimate the future value of the response variable. In digital library this is a very useful technique to predict future trends and rate of absorption of a new idea.

4.1.6 Abnormality Detection:

This technique is used to identify outliers and extreme data behavior. Such anomalous characteristics have been in the past a great sources of new theories and discoveries. Digital Libraries can use this technique to guide frontier research.

5. Application of Text / Data Mining

The prime aim of the text / data mining is to identify the useful information without duplication from various documents with synonymous understanding. TDM is an empirical tool that has a capacity of identifying new information that is not apparent from a document collection. The primary focus of text mining is classification and prediction. These are also the methods widely used in TDM. The idea is, given a set of past experiences along with correct answers, the objective is to find an answer to a new experience. This type of methods is called text categorization. This is one of the important applications of TDM.

6. Document Classification

It most widely used application of text mining. Text categorization groups the heterogeneous collection of documents into homogeneous groups based on their similarity. When a new document is presented, it is matched against profile of each of the groups and assigned to a folder appropriately. This is shown schematically in Fig.6

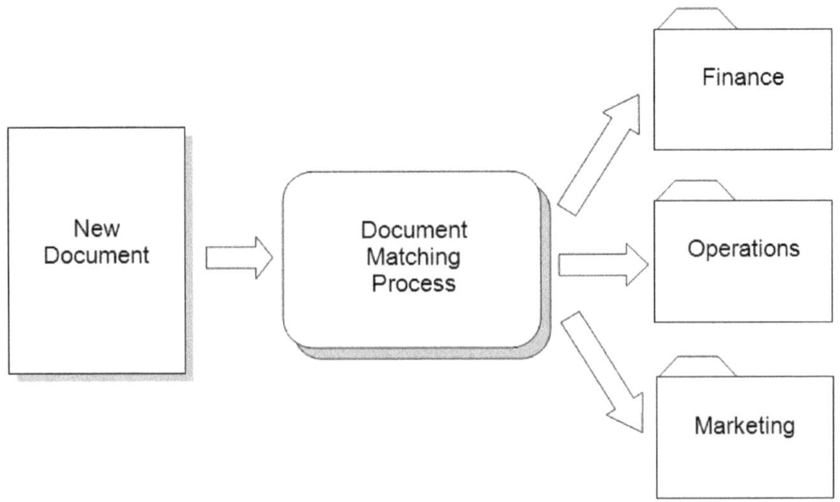

Source: Weiss, Sholom M et al, Ed (2005), Text Mining, Predictive Methods for Analyzing Unstructured Information, Springer

Fig.6 Document Classification process

7. Information Retrieval

IR is a topic more associated with online documents and search engines. On the net there is a vast collection of documents. Only one or a few of them are of interest to any one individual. The goal of IR is to retrieve those documents that satisfy the user's need. So the input to the IR system is the clues that specify the user need. These clues are called keywords or a query. The basic concept of IR system is measuring similarity. A comparison is made as to how similar two documents are. Similar to text categorization method, the IR process transforms the user query into a document, matches the query document to each of the documents in the collection and retrieves the matched documents from the collection. The IR system is schematically shown in Fig. 7.

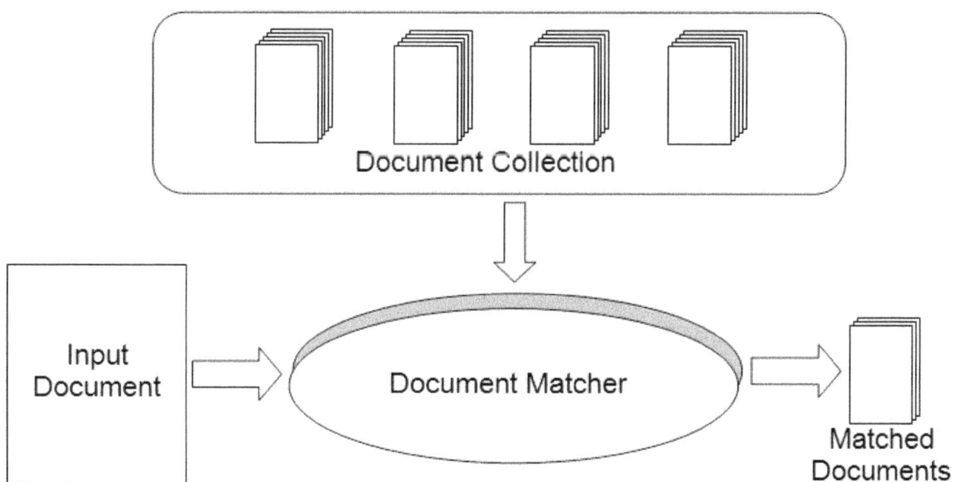

Source: Weiss, Sholom M et al, Ed (2005), Text Mining, Predictive Methods for Analyzing Unstructured Information, Springer

Fig. 7 IR System Process

8. Clustering

In document classification, the goal was to organize text documents into folders appropriately based on their subject matter. These folders might have been pre-created and decision criteria to decide which document goes to which folder might have been specified by a subject expert. One can imagine a situation where the folders are not created before-hand and the required expertise is not available. Clustering is a technique where human supervision is not required to complete these tasks before its use. For this reason, clustering is referred to as unsupervised machine learning technique. Clustering considers each document as a point or a vector in an "n" dimensional space and analyzes the distance between the documents to establish similarity and arrange them into folders. The clustering process is shown schematically in Fig. 8

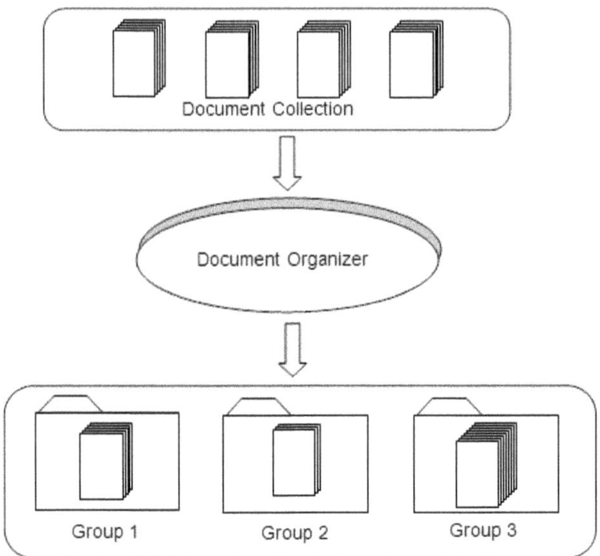

Source: Weiss, Sholom M. et al, Ed (2005), Text Mining, Predictive Methods for Analyzing Unstructured Information, Springer

Fig.8 Clustering Process

9. Information extraction

The techniques so far explained used basic and shallow measurements such as presence or absence of words to categorize text documents. There is definitely much more information in the document which is not captured by these methods. It is indeed a fine proposition to extract this information and make it compatible to more advanced statistical or data mining techniques to learn much more from textual data. This type of effort is called information extraction and is schematically shown in Fig.9

Director's note to Annual Statement for YE 2011-12

... achieved a revenue of fifty two million dollars and reported a profit of 4.5 million for this fiscal year....

Spread Sheet

...	Revenue	...	Profit	...
...	...	...	...	...
...	...	...	...	...
...	...	...	...	...
...	52000000	...	450000	...
...	...	...	...	...
...	...	...	...	...

Source:. Weiss, Sholom M et al, Ed (2005), Text Mining, Predictive Methods for Analyzing Unstructured Information, Springer

Fig.9 .Information Extraction Process

The applications of text /data mining include the following:

- text categorization into specific domains for example spam - non spam emails or for detecting sexually explicit content ;
- text clustering to automatically organize a set of documents.
- sentiment analysis to identify and extract subjective information in documents. Detect what your customers are saying about your company when they use social media
- concept/entity extraction that is capable of identifying people, places, organizations, and other entities from documents.
- document summarization to automatically provide the most important points in the original document.
- learning relations between named entities.
- Spam filtering
- Creating suggestion and recommendations (like Amazon)
- Monitoring public opinions (for example in blogs or review sites)
- Customer service, email support
- Automatic labeling of documents in business libraries
- Measuring customer preferences by analyzing qualitative interviews
- Fraud detection by investigating notification of claims
- Fighting cyber bullying or cybercrime in IM and IRC chat

10. Key challenges of Text /Data Mining Application

TM also known as Text Data Mining (TDM) or KDT refers generally to the process of extracting interesting and non-trivial information and knowledge from unstructured text. Text mining is an interdisciplinary field which draws on information retrieval, data mining, machine learning, statistics and computational linguistics (Vidhya and Aghila 2010). The following key challenges of Text / Data Mining can be stated:

10.1 Computer network support:

The data mining and information resource construction in the digital library has to be on a powerful and robust computer network. Information transfer rates and storage capacity should be adequate not only for today but for years to come as the library resources have an explosive tendency to grow. It is convenient to construct large-scale data warehouse, which caters to scalability of text mining operations. Since text mining is collection of technologies and also is emerging, a single tool or method is not able to cater to all the end-to-end requirements. So a cross functional platform with interoperable software integration design must be adopted.

10.2 Depth mining:

Since Text Mining is concerned with hidden pattern and obscure knowledge an in-depth mass data mining is extremely important. Mathematical, statistical methods, regression analysis, association analysis and artificial intelligence may solve some of depth mining tasks. Data Analysts required for information mining must possess different level of skills compared to general information analysts. These analysts must have a good grasp of computer science, statistical methods and database and data warehouse technologies.

11. Role of Library Professionals towards Text/ Data Mining

"Necessity is the mother of Invention" as the saying goes. Accordingly the library community has found solution to the scholarly communication crisis in the form of Open Access sources and other forms of information including unstructured data through TDM and need for imparting Text / data mining literacy. The new generation information professionals have to be aware of the revolution and evolutions taking place in text/data mining and its influence on the access to information. The LIS professionals shall equip themselves with the necessary ICT skills and competencies including TDM techniques and methodologies in the use of the emerging open access resources including databases and a large amount of unstructured data available in the corporate sector. They shall facilitate the researchers and other academics in the art of self-archiving to such contents. They can also help the academics for proxy self-archiving on behalf of the others who are unable to self-archive. Further they can collaborate with IT system administrators to ensure the upkeep of the systems, backup, monitoring, upgrading of the institutional open archives (Ramesh Babu, 2013, 2014).

There are a number of additional things that librarians can do to assist with Text/Data Mining (TDM), which includes the following:

- Highlighting the relevance and application of the TDM exception as part of any copyright training they offer to researchers and advertising TDM support as a service. This could be promoted alongside related research support services.
- Encouraging the development of partnerships with academic colleagues to work on TDM projects, through highlighting institutional collections that might be suitable for mining.
- Being clear about licensing and terms & conditions under which resources are made available to colleagues at your institution, and ensuring they fully understand the copyright exception that permits TDM for non-commercial research regardless of contractual stipulations.
- Being firm with publishers and other content suppliers in the case of 'unusual behaviour' reports or Digital Rights Management (DRM) blocks to protect the legitimate interests of researchers, as defined by law.
- Advising on next steps should local support not be enough, up to and including appealing to the Intellectual Property Office (IPO) if a rights-holder will not remove DRM locks. Information on making a complaint can be found on the IPO's website.
- Being aware of data protection issues: TDM is not exempted from data protection law, so content identifying a living individual must be processed in compliance with the Data Protection Act (1998).
- Being mindful of clauses in any new license agreements for resources that might restrict TDM activities (Secker, *et al*, 2016).

The role of Information Professionals has been changing over the years to align it with the changing needs of the users. For example, datasets were created in order to store existing information, and Information Professionals are responsible for retrieving that information for future usage by the users/researchers. On the other hand, TDM emphasises on generating new information from existing one, which means mining through the existing databases to create unknown information in order to produce new products that are currently not existing (Mogashoa, 2015). Therefore, library professionals have to act in a proactive manner to support the twenty-first century educational change. Professionals working in such libraries need continuous grooming by acquiring core competencies and new skills so that they never become obsolete in this fast changing environment. In other words, to work efficiently and effectively in the fast-changing digital age, a new generation of LIS professionals should have the skills and competencies including text /data mining techniques in providing information. Without possessing or acquiring those skills and techniques, the LIS professionals will not be able to fulfil their role in any Information centre. Unless you have these TDM skills in addition to other skills, you cannot get to the competencies.

12. Conclusion

The world is experiencing an explosion in the availability of information and has become information driven society. Individual and organization success is characterized by effective and efficient processing of this information to derive actionable knowledge. Technology is empowering this transformation by providing sophisticated tools for processing information. Information explosion and availability of information in various forms has changed the shape of information centres and nature of information profession and professionals. Information profession and professionals have been impacted by the exponentially increasing volumes of information available - as well as with changing attitudes and behaviour of information seeker toward electronic resources. It is very difficult to find the required pieces of information in the bundled of scattered information. Here text / data mining have come as a tool to help Information professionals to find the relevant information and deliver to its users. Text mining is used as a technology for analyzing large volumes of structured/unstructured textual documents. Text mining has very high knowledge and commercial values.

The aim of Text / Data mining is generally to strengthen decision making and internal operations processes of any organisation and generation of new domain of knowledge. These technologies help to increase the utilization of Knowledge Management (KM) systems and pro-actively help information professionals to improve their competencies and thus productivity of the organization (Verma, Ranjan and Priyanka Mishra 2015). TDM is an inter-disciplinary field that combines methodologies from various other areas such as Information Retrieval (IR), Information Extraction (IE), and utilizes techniques from the general field of data mining, e.g., clustering. It mining is the core component of knowledge discovery and management process. It a fact that more than eighty percent of information today exists in format free textual form. It offers opportunity to the libraries to improve their services effectively. TDM technology can help in deep analysis of information, extraction of inter linkages and mass customization of individual processing needs to promote the digital library. Librarians can use the principles of TDM to achieve the organizational goals. It can be incorporated into many library operations to improve not only their effectiveness but also strengthening of relationships with related units inside and outside the institution. Ultimately, librarians should be mindful of the right to mine, and recognise the important role they can play in liberating data and knowledge. Their continuing mission: to explore strange new worlds, to support research and discovery of new scientific breakthroughs, to boldly go where no one has gone before. TDM is an emerging field in data analysis where researchers are not only developing new techniques for knowledge discovery but also are trying to apply techniques from related disciplines like statistics and data mining. It is now an emerging area within which many powerful methods of text analysis and knowledge exists. The field is still in a fledgling state and provides immense opportunities for future research.

References

Alpaydin, E. (2004). *Introduction to Machine Learning*. New Delhi, PHI Learning.

Berry, Michael W and Castellanos, Malu (2007). *Survey of Text Mining: Clustering, Classification, and Retrieval*, 2nd ed., Springer.

Chandrasekhar, C K and Ramesh Babu, B (2012) Text Mining: Tool for Enhancing Fact-based Managerial Informed Decision Making, *International Conference on Business Intelligence, Analytics and Knowledge Management*, IBS, Hyderabad.

Chandrasekhar, C. K. (2012). Application of Text Mining as Content Management Tool in Digital Libraries. IN: *Dynamics of Librarianship in the Knowledge Society. (Festschrift in Honour of Prof. B. Ramesh Babu), Edited by Achim Osswald et al., Vol. 2.* New Delhi, B.R. Publishing Corporation, pp. 770-778.

Chandrasekhar, C. K. and Ramesh Babu, B. (2013). Application of Text Mining as a Content Management Tool in Digital Libraries. *New Dimensions in Web Based Library and Information services. Edited by Saroja, G. and Chandrasehara Rao, V.,* New Delhi, Pearl Books, pp. 25-34.

Chandrasekhar, C.K; Ramesh Babu, B and Srinivasan, M.R (2016). Integrating Data mining Techniques into Digital Library services. IN: *Marching Beyond Libraries : Role of Social Media and Networking* (National Conference Proceedings), edited by Bijayalakshmi Rautary, Dilip K Swain and Chandrakant Swain. Bhubaneswar& New Delhi: KIIT University and Overseas Press India Pvt. Ltd, pp. 126-135.

Chandrasekhar, CK and Ramesh Babu, B (2012a). Empowering Next generation Libraries through Text Mining: TM a content management tool for the new millennium. IN: *Emerging Trends in User Expectations for Next Generation Users*, Edited by M. Dorasamy, B. Ramesh Babu and Raavi Sarada. Vijayawada, Andhra Pradesh Library Association, 2012 pp. 24-32

Chowdhury, G.G. (1999) *Introduction to Modern Information Retrieval*. London: Library Association.

Clark, Jonathan (2013). *Text Mining and Scholarly Publishing, a report for the Publishing Research Consortium*. Loosdrecht, The Netherlands & London, pp. 5-6.

Deepika, S. and Mehta, S. (2014). A Study of Data Mining Clustering Techniques. *International Journal of Advanced Research in Computer Science and Software Engineering,* 4(3): 490-494.

Grimes, S. (2007). A Brief History of Text Analytics. *Beyenetwork.* Available at: http://www.b-eye-network.com/view/6311. (Accessed on 25th June 2017).

Hand D.J., Mannila H., and Smyth P. (2001) *Principles of data mining,* MIT Press

Hearst, Marti A. (1999) Untangling Text Data Mining, *Proceedings of ACL'99: the 37th Annual Meeting of the Association for Computational Linguistics,* University of Maryland, June 20-26, 1999. (Available at http://people.ischool.berkeley.edu/~hearst/papers/acl99/acl99-tdm.html) Accessed on 21 June 2017.

IFLA Statement on Text and Data Mining (2013). Available at https://www.ifla.org/publications/ifla-statement-on-text-and-data-mining-2013 (Accessed on 23-06-2017)

Jiawei Han and Micheline Kamber (2000). *Data Mining: Concepts and Techniques,* The Morgan Kaufmann Publishers.

Kanya, N. and Geetha, S. (2007). Information Extraction: A Text Mining Approach. *IET-UK International Conference on Information and Communication Technology in Electrical Sciences, IEEE,* Chennai, pp. 1111-1118.

Liritano, S, and Ruffolo, M, (2001). Managing the Knowledge Contained in Electronic Documents: A Clustering Method for Text Mining", *IEEE,* Italy pp.454-458

Mogashoa, Daniel (2015). Text Mining for Librarians/Information Professionals. Available at https://www.linkedin.com/pulse/text-mining-librariansinformation-professionals-daniel-mogashoa (Accessed on 23-06-2017)

Pyle, D (1999). *Data Preparation for Data Mining,* 2nd ed., Morgan Kaufmann.

Ramesh Babu, B (2011). Content Management Stratégies, Tools and technologies: An overview. In: *Library and Information Services in the Digital Era,* edited by L S Ramaiah, et al , Hyderabad: BS Publications, pp.328-343

Ramesh Babu, B (2012). Information Literacy Skills for LIS Professionals. IN: *Library Services in the Knowledge Web* (Collection of papers in honour of Dr Stanley Madan Kumar) edited by K Veeranjaneyulu *et al.* New Delhi: New India publishing Agency. pp.214- 223

Ramesh Babu, B (2014). Digital Information Literacy: An Overview. IN: *Trends in Library and Information Science* (Prof. A Manoharan festschrift), edited by K S Sivakumaren et al, Chennai : Almighty Book Company, pp. 257-263.

Ramesh Babu, B (2014a). Knowledge management in electronic environment: Opportunities and Challenges for LIS professionals. (Keynote address) Proceedings of the UGC sponsored *National Seminar on Knowledge Management in Electronic Environment : Opportunities and Challenges* organized during *21-22 March 2014,* edited by R Jeyshankar and S Thanuskodi. Karaikkudi : Alagappa University, Department of Library and Information Science, Pp. 1-12

Ramesh Babu, B et al, Ed. (2003). *Knowledge Management: Today and Tomorrow.* (Festschrift in honour of Dr. Velaga Venkatappaiah). New Delhi: Ess Ess Publications.

Ramesh Babu, B *et* al, Eds. (2004). *Information Management: Trends and Issues.* (Prof.S.Seetharama Festschrift) New Delhi: Researchco Book Centre.

Sebastiani, F. (2002) Machine learning in automated text categorization. *ACM Computing Surveys,* 34 (1): 1–47.

Secker, Jane *et al*, (2016). To boldly go… the librarian's role in text and data mining. Available at https://www.cilip.org.uk/blog/boldly-go-librarians-role-text-data-mining (Accessed on 23-06-2017)

Solka, J.L. (2008). Text Data Mining: Theory and Methods. *Project Euclid Mathematics and Statistics online: Statistics Survey*, 2: 94-112.

Verma, Vijay Kumar; Ranjan, Manish and Priyanka Mishra (2015). Text mining and information professionals: Role, issues and challenges. IN: *4th International Symposium on Emerging Trends and Technologies in Libraries and Information Services (ETTLIS)* held during 6-8 January, 2015, Institute of Electrical and Electronics Engineers, Inc, pp. 133-138

Vidhya. K. A and Aghila G (2010). Text Mining Process, Techniques and Tools: an Overview. *International Journal of Information Technology and Knowledge Management*, 2 (2): 613-622

Vijayarani, S; Ilamathi, J and Nithya (2015). Preprocessing Techniques for Text Mining - An Overview . *International Journal of Computer Science & Communication Networks*, 5(1):7-16

Weiss S. M. *et al*. Eds. (2005). *Text Mining: Predictive Methods for Analyzing Unstructured Information*. New York, Springer

Witten, I.H (2005). Text Mining. In: *Practical handbook of internet computing*, edited by M.P. Singh, pp. 14-1 - 14-22. Chapman & Hall/CRC Press, Boca Raton, Florida. Available at www.cos.ufrj.br/~jano/LinkedDocuments/_papers/aula13/04-IHW-Textmining.pdf (Accessed on 19th June 2017)

Witten, I.H. and Frank, E. (2000) *Data mining: Practical machine learning tools and techniques with Java implementations*. Morgan Kaufmann, San Francisco, CA

Zhang, Mei (2011). Application of Data Mining Technology in Digital Library, *Journal of Computers* 6(4).

2

Virtual Teams: Creating Professional Communities

Dr. Stanley Madan Kumar

University Librarian (Retd)
University of Agricultural Sciences, Dharwad-580005, Karnataka
e-mail: sunnystanmk@gmail.com

ABSTRACT

Virtual Teams (VT) are unconventional social configurations highly depended on ICT and are not constrained by spatial and temporal boundaries. Advances in ICT have created new opportunities for introducing innovative library services suitable to the dynamic digital era. With the increasing impact of ICT and digital technology, reference and information services have increasingly moved away from library reference desks out into the virtual world. The present paper deals with the causes for increasing virtual environment, demand for virtual library information services, need for user education and continuing education of librarians and the future of library reference services which at present are passing through a transitory phase.

Key Words: *Virtual teams, computer mediated communication, communication cohesion, library information services*

1. Introduction

Virtual team (VT) is described as a small number of people with complementary skills who are equally committed to a common purpose, goals and working approach for which they hold themselves mutually accountable.

The VT are often formed to overcome spatial separation. Advances in ICT have created new opportunities for creating VT. Such teams comprise of employees with unique skills, located at different places and collaborate with each other to accomplish designated tasks. VT members may be located across a country or across globe, rarely meet face to face and include members from different cultures (Kirkman et al, 2002).

2. Causes for VT Formation

Economic activities of all types are moving in the direction of globalization. With the rapid development of electronic information and communication media in the past decades, distributed work has become much easier, faster and more efficient. IT is providing the infrastructure necessary to support the development of new organizational forms –and VT is one such organizational form (Powell et al, 2004).

Summarizing, a VT can be described as –

- Geographically dispersed
- Driven by a common purpose
- Predominantly work with ICT
- Involved in cross boundary collaborations
- Team members are knowledge workers
- Small team size
- Not a permanent team
- Members may belong to different organizations and cultures

3. Success Factors of VT

The success of VT is dependent on the expertise, knowledge and wisdom of individuals, time management and self-supervisory skills. Team selection is a crucial factor. In today's competitive global economy, organizations capable of creating VT talented people can respond quickly to changing environment (Bergiel et al. 2008) Working of the VT involves exchange of data through internet

4. Advantages and Disadvantages of VT

With the rapid development of electronic information and communication media, distributed work has become easier, faster and more efficient (Hertel et al. 2005). In lieu of face to face contact dispersed teams have to collaborate and share their knowledge through an integrated set of tools that enables complex communication and serves as a common information repository that is easily revised and accessible 24x7, any place (Lipnack and Stamps, 1997).

While there are great advantages that come with the adoption of the VTs, new challenges rise with them. Cascio (2000) identified 5 main disadvantages: lack of physical interaction, loss of face-to-face synergies; lack of trust; greater concern with predictability and reliability, and lack of social interaction.

5. Characteristics of VT Members

Relationships are critical for efficient organizational work. The changes in work culture affect the way VT members conduct their work and how they interact..

- Members have to learn new ways to express themselves and to understand others
- Must have superior team participation skills
- Must be experts in a variety of computer-based technologies. This requires additional developments in areas of communication and cultural diversity
- Membership will cross national boundaries and cultural backgrounds. This complicates communication and work interaction (Townsend et al, 1999)

6. Virtuality in Library Services

VT evolve in the field of library services when the libraries share collection, inter-library lending, collective bargaining for consortial subscription packages, online access, share metadata and records, jointly organize e-learning courses and continuing education programs. There are three levels of virtuality in libraries (Bauwens, 1994).

1. Electronic access (OPAC)
2. E-access virtual collections combined to document delivery
3. E–access to virtual collections which are contributed by e-documents

A hybrid-digital environment provide libraries with significant opportunities to play. They could be places where people could come and browse carefully created collections and explore subject areas they might not know existed and would like to know more about. Librarians can take advantage of networked reference service models for virtual reference to build high quality, highly efficient reference service.

Librarians occupy a unique intermediary role between the role of supplier and consumer. They provide wide range of information support services not only to their patrons of parent organization's community. They serve as intermediary between other campus consumers and content providers. Further, they need to have wide ranging access to the entire collection of materials. Research collections are expensive to acquire, develop, expand and service. They require extensive and expensive preservation and conservation work (Al Rashid,1998).

7. Changing User Habits

Today's digital age has brought changes to many aspects of library services. Users now search for information from their desktops and laptops, download e-books on their laptops. Full text retrieval of information source is becoming common place and information services are increasingly becoming personalized.

The power of Internet and mobile technology has enabled us to live with and access an incredible range of data, information and services. Today, just 'Google it' seems to be everyone's answer for information. VT is based on the infrastructure provided by the Internet and the web (Smith, 1993). It is still in the stage of continuous and rapid evolution and development. To participate in and develop virtual space for information handling and management, today's librarians require specific skills.

8. Impact on Reference Services

With the increasing influence of online information and information environment, reference and information services have increasingly moved away from library reference desks and away from libraries' print collection out into the virtual world. This is more due to the increasing influence of online environment, electronic information explosion and a steady stream of new innovative information technologies being developed.

The NetGen individuals also described as 'digital natives' often pose a challenge to librarians as they belong to networked, interactive and multimedia-oriented culture. Easy access to information via web has changed their concept of information environment. They are not only consumers of information, also active information creators (Stanley,2015). Information market is fast developing into a multimillion dollar enterprise. Library users have become information consumers. Profit-oriented publishing houses are selling information directly to scholars and students (D'Andria, 1994).

The amount and diversity of the ever increasing information on the Internet and in online data bases is one of the major attributes to the increased role of library information service units (Mayega,2008) With the increase in the availability of information on computers, laptops, smart phones etc. users prefer 'clicking a button' to actually turning the pages of a book. They are used to information that is concise, instant and fits within their schedules. Wikipedia is preferred over other conventional encyclopedias like Britannica.

9. Traditional Skills Vs. Digital Technology

Extensive use of computers, increased reliance on networks, rapid growth of Internet and information explosion have compelled librarians to adapt new and innovative means and methods for assessing storage, retrieval and dissemination of information. Internet is no more a "browse and serf" environment.

In addition, they have to seek affiliation with similar institutions to survive in the changing environment. Traditionally they had these facilities in the form of inter-library lending and collection development. These alliances would establish formidable collections of resources, considerable convenience to information users and significantly strengthen the position of academic libraries in information market place (Skiadas,1999).

10. LIS Education

In this context of digital spaced and dynamic developments of virtual forms of products and services much attention should be paid towards to LIS teaching so that new skills and incorporated in the curricula. They refer to the ways, methods, tools and standards to be used to handle the dynamic information. Since the digital environment is in a stage of continuous evolution, this also is the case for the LIS curricula.

Information professionals should be able to cope with both traditional skills of handling and managing information products, services and institutions, and with the new requirements concerning tools, methods and techniques for handling information and information services in the digital environment. Librarians have to ensure that proper training facilities including courses are available for the users as well as library staff. From this point of view much attention should be paid by LIS schools so that necessary skills are incorporated in the relevant curricula.

11. Conclusion

Virtual Team represents a new form of organization that offers unprecedented levels of flexibility and responsiveness and has the potential to revolutionize the workplace (Powell,et al, 2004). VT allows people to collaborate more productively at a distance. However, management experts argue that VT are not appropriate for all circumstances. Virtual proximity, connectivity facilitated by the use of ICT cannot completely substitute the physical proximity when it comes to innovation and learning. But, is still the most reliable and effective way to review and revise new ideas.

The interaction in computer-mediated, communication environment is more impersonal, more task-oriented, more businesslike and less friendly than in face-to-face settings (Schmidt et al 2011). ICT enabled virtual collaboration would be effective with the existence of face-to-face communication support and would lead to higher levels of satisfaction in collaboration.

References

Al Rashid,1998. Safeguarding copyrighted contents: Digital libraries and intellectual property management. Available at http.www.dlib.org/dlib/april98/04.barker.html

Bauwens (M), 1994. The role of cybrarians in emerging virtual age. FID News Bulletin 44(7/8):131-137

Bergiel (JB), Bergiel (EB) and Baismeir (PW), 2008. Nature of virtual teams: A summary of their advantages ad disadvantages. Management Research News 31:99-110

Cascio (WF) and Sharygaibo(S), 2003.E-leadership and virtual teams. Organizational Dynamics 31:363-376

Cascio (WF),2000. Managing virtual work place. The Academy of Management Executive 14:81-90

Conroy (B),1983. People networks: A system for library change. Journal of Library Administration 4(2):75-86

D'Andria (Fr. A),1994. The business of libraries is staying in business. Journal of Library Administration 20(2):81-91

Hertel (GT),Geister (S) and Konrad (U), 2005. Managing virtual teams. A review of current research. Human Resource Management Review 15:69-95

Kirkman (BL), Rosen (B), Gibson (CB), Tesluk (PE) and McPherson (S),2002. Five challenges to virtual teams success: Lessons from Sabre Inc. Academy of Management Executive 16(3): 67-79

Lipnack (J) and Stamps (J), 1997. Virtual teams: Reaching across space, time and organizations, NY, Wiley.

Peters (LM) and Manz (CC),2007. Identify antecedents of virtual team collaboration. Team Performance Management 13: 117-129

Powell (A), Piccolli (G) and Ives (B), 2004. Virtual teams: A review of current literature and directions for future research. The Database for Advances in Information Systems 35: 6-36

Schmidt (JM), Montoya-Weiss (MM) and Massey (AP), 2001. New Product development decision making effectiveness : Comparing Individuals face-to-face teams and virtual teams. Decision Science 32:1-26

Skiadas (Ch),1999. The role of libraries in a changing academic environment. 20th IATUL Conference Proceedings (17-21 May) (Technical University of Crete, Greece)

Smith (NR), 1993. The Golden Triangle – Users, Librarian and Suppliers in the electronic information era. Information Science and Use 13: 17-24

Stanley (MK) ,2014. Role of libraries and information centers in the 21st century (Sambhram Group of Institutions and KALA, Bangalore)(Jan).

Stanley (MK), 2015. Virtual teams and library profession: Bridging space over time . National Conference on Knowledge Discovery and Management (NCKDM 2015) (Calicut University, Calicut)(30- 31 Jan)

3

Awareness of Agricultural Information Systems Among the Postgraduate Research Scholars of Prof.Jayashankar Telangana State Agricultural University, Hyderabad: A Case Study

Dr. K. Veeranjaneyulu

University Librarian & Professor and
Professor Jayashankar Telangana State Agricultural University
Hyderabad – 500 030, Telengana State
e-mail: veeru030463@gmail.com

ABSTRACT

A study has been undertaken to know the awareness, satisfaction level, purpose and problems faced by the postgraduate research scholars of Professor Jayashankar Telengana State Agricultural University (PJTSAU) while accessing agricultural information systems (AIS). The study has been conducted using a questionnaire with 217 and 101 research scholars undergoing Master and Doctoral research programmes at PJTSAU, Hyderabad. It is found that 100 per cent of the respondents are aware about the AIS. It has been found that 74% of research scholars are using AIS for research purpose, 52 % for educational purpose and followed by current information (16%), to update Knowledge (12%) and writing papers for publication (8%). It is also noted that majority of students are not facing problems while seeking information from Agricultural Information Systems. This highlights the findings of the study and also suggested to conduct regular awareness user awareness programmes, provide high bandwidth Internet connectivity and to provide more e-resources.

Keywords: *Agricultural Information Systems, Agricultural Library, PJTSAU, Information Literacy, User Study, CeRA, Krishikosh, Agricat, ICAR, NAIP*

Introduction

The University Library, Professor Jayashankar Telengana State Agricultural University (PJTSAU) has very rich collections of print and non-print documents viz., of books, E-books, e-journals, Databases, Krishikosh etc. All the resources are made available through offline and online. The University Library is catering the needs of teachers, scientist, extension specialist, research scholars and students. PJTSAU has implemented many schemes supported by the Indian Council of Agricultural Research (ICAR), New Delhi and implemented many state-of-the-art facilities in the Library. The Library is considered to be one of the best agricultural libraries in the country on the basis of its collections and services.

Purpose of the Study

The purpose of the study is to know the awareness, satisfaction level, purpose and problems faced by the postgraduate research scholars while accessing Agricultural Information Systems.

Objectives of the Study

The objectives of the study are as follow:

1. To study the awareness of postgraduate research scholars about Agricultural Information Systems.
2. To know the purpose of using Agricultural Information Systems.
3. To study the satisfaction level of research scholars regarding Agricultural Information Systems.
4. Problems faced by the postgraduate research scholars while seeking information through Agricultural Information Systems.

Limitations of the Study

The study is limited to postgraduate research scholars (P.G. & Ph.D.) students of Professor Jayashankar Telangana State Agricultural University. Agricultural Information Systems covered the study are CeRA, Krishikosh and Agricat.

Methodology

The Data required for the study was collected from primary as well as secondary sources. The primary data was collected through questionnaires from Postgraduate students. For data collection 125 questionnaires were distributed to the students of

PJTS Agricultural University. It is about 40% of the population surveyed. 80% of the population in this study responded to our questionnaire which sufficient fair percentage to deduce the inferences for the study

Analysis of Data

Data collected from postgraduate research scholars regarding Agricultural Information Systems are analyzed below:

Distribution of Sampled Respondents based on Gender

Table No.1 Distribution of Students based on Gender and Course-wise

Sl. No.	Course	Gender		Total
		Male	Female	(%)
1.	P.G.	96	121	217 (68)
2.	Ph.D.	63	38	101 (32)
Total		**159**	**159**	**318 (100)**

The above table shows that 318 students are study postgraduate at PJTS Agricultural University. Out of which 217 are P.G. and 101 are Ph. D. Students respectively. And also reveals that the number of male and female students enrolling into P.G. and Ph.D., which shows that there is no variation of gender rate and it is interesting to note that both the male and female are equal i.e. 159 and 159 respectively.

Awareness levels of the Agricultural Information Systems

Table No. 2 Awareness about Agricultural Information Systems

Sl. No.	Awareness	No. of Respondents	%
1.	Aware	100	100
2.	Not Aware	--	--
Total		**100**	**100**

The table no. 2 shows that 100 percent of the respondents are aware about the Agricultural Information Systems. Respondents expressed that they have completed course 501 – Library Science. Therefore, all the students are aware of Agricultural Information Systems.

Purpose of using E-Resources

Table No. 3 Purpose of using Agricultural Information Systems

Sl. No.	Purpose	Total Respondents	No. of Respondents	%
1.	Education	100	52	52
2.	Research	100	74	74
3.	Current Information	100	16	16
4.	Writing papers for publication	100	8	8
5.	To update Knowledge	100	12	12

Table No. 3 explains that 74% of students are using Agricultural Information Systems for research purpose, 52 % for educational purpose and followed by current information (16%), to update Knowledge (12%) and Writing papers for publication (8%). Majority of the students are using Agricultural Information Systems for research and educational purpose

Satisfaction Levels among the Students

Table No. 4 Satisfaction level of students regarding Agricultural Information Systems

Sl. No.	Resources	Excellent	Good	Fair	Poor	Total
1.	CeRA	56	28	14	2	100
2.	Krishikosh	15	36	41	8	100
3.	Agricat	14	22	42	22	100

The above table shows that majority (56%) of students are satisfied with CeRA resources and expressed the satisfied level is excellent. 41% and 42% respondents expressed that the level of satisfaction of Krishikosh and Agricat respectively are fair. The Table No. 4 reveals that majority of students are using CeRA.

Problems faced by the Students while seeking information

Table No. 5 Problems faced by the Students while seeking information

Sl. No.	Problems	Problems faced	Problems not faced	Total
1.	Non-availability of materials	26	73	100
2.	Inadequate services of library staff	6	94	100
3.	Unsuitable timings of the library	11	89	100
4.	Lack of Knowledge in using the library	10	90	100
5.	Scattered information in too many sources	19	81	100
6.	Outdated information	12	88	100

Majority of students are not facing problems while seeking information from Agricultural Information Systems. Non-availability of materials and scattered information in too many sources are the major problems faced by the students. To overcome the problems regular user awareness programs should be conducted

Findings

On the analysis of questionnaire, the following are findings of the study.

1. 100 percent of the respondents are aware about the Agricultural Information Systems
2. Majority of the students are using Agricultural Information Systems for research and educational purpose.

3. 56% of students are satisfied with CeRA resources and expressed the satisfied level is excellent. 41% and 42% respondents expressed that the level of satisfaction of Krishikosh and Agricat respectively are fair.
4. Majority of students are not facing problems while seeking information from Agricultural Information Systems.

Suggestions

In the light of above findings, the following suggestions are made:

1. Regular user awareness user awareness training programmes may be conducted for students and staff.
2. High bandwidth Internet facility should be provided to provide uninterrupted services.
3. To overcome the problems faced by students, library staff should guide while searching information from Agricultural Information Systems.
4. More e-Resources should be included to canter the needs of the users.

References

Jain, Arun Kumar and Veeranjaneyulu, K (2014). Repository of Indian National Agricultural Research & Education System (NARS) : Open Access to Institutional Knowledge. Indian Journal of Agricultural Library and Information Services 30 (1), pp. 1-5.

Nirmal Singh (2014). Consortium for e-resources in agriculture: Qualitative and quantitative perspectives. Current Science 107 (7), pp. 1112-1117.

Rathinasabapathy, G.; Veeranjaneyulu, K. and Amarendar Kumar (2016). Krishikosh : Institutional Repository of National Agricultural Education and Research System (NARES) of India : An Analytical Study. ICDL 2016 - Smart Future : Knowledge Trends that will change the world, New Delhi. pp. 873-887. ISBN: 9788179936535

Selvaraj, A. D., and G. Rathinasabapathy (2016). A Study on Electronic Information Use Pattern of Faculty Members of Self-Financing Engineering Colleges in Tiruvallur District, Tamil Nadu. Asian Journal of Library and Information Science 6.3-4 (2016): 31-40.

Selvaraj, A. D., and G. Rathinasabapathy. (2015). Information Use Pattern by Students of Self-Financing Engineering Colleges in Tiruvallur District, Tamil Nadu: A Study. Journal of Library, Information and Communication Technology 6.3-4 (2015): 45-54.

Veeranjaneyulu, K. (2013). KrishiKosh: An Institutional Repository of National Agricultural Research System in India. Library Management 35 (4&5), pp.345-354.

Veeranjaneyulu, K. and Ravi Kumar, N.P.(2015). Consortium for e-Resources in Agriculture (CeRA) : A great gateway to e-Journals. Indian Journal of Information, Library & Society 28 (1-2), pp. 110-116.

Veeranjaneyulu, K. Ravi Kumar, N.P (2011). Impact of Digital Environment on Academic Library. Human Resources Transformation of Agricultural Libraries in Collaborative Era, Hyderabad : BS Publications.

4

Krishikosh: A Digital Repository for Knowledge Discovery

Amrender Kumar and Rakhee Sharma

Agricultural Knowledge Management Unit,
ICAR-Indian Agricultural Research Institute
Pusa, New Delhi – 110 012
e-mail: akjha@iari.res.in

ABSTRACT

Krishikosh is a versatile open access digital repository catering to the needs of Indian National Agricultural Research and Education System (NARES) and has architecture of centralized hosting of content but decentralized management. The Krishikosh has been designed using open source software DSpace and an efficient Integrated Content Management System (ICMS), which has suitability configured to meet the requirements of NARES and created dependable digital repository. Each institution in NARES has been configured as community in DSpace having it's own collections and logo. Each community and collection can be given independent rights to registered users for uploading and managing the contents. Effectively Krishikosh is a collectively managed, centrally aggregated repository with integrated search facility. Krishikosh is a huge repository of thesis, reports, articles etc. Keeping track of all the information is very hectic and difficult for the user, which results in low usage of the Krishikosh site. Thus, an application with push notification was developed in order to involve the user with the Krishikosh, by sending messages regarding the new upload. Google Cloud Messaging (GCM) server were utilized for implementation of push notification which take the data from the postgres SQL database, it is the most advanced open source database. PHP script is used to connect the java programming in android studio, the database and the GCM server. In this study, usage statistics of Krishikosh were also analyzed.

Keywords: *Institutional Repository, KrishiKosh, e-Granth, ICAR, NAIP, NARES, Integrated Content Management System*

1. Introduction

Right information at right time is crucial for any development. Agriculture is no exception. Access to right and timely information is as important input as seeds, soil, water, pesticides and other inputs for any successful farm operation. Today, quick access to authentic information has become absolutely important to optimize the agricultural output per unit land/inputs, mitigate the effects of abrupt climate changes, sustainability of natural resources and ensuring quality for nutritional security. Therefore, the demand for fast access to authentic and credible digital information sources have risen in agriculture sector be it research, education or extension. End to end value chain development requires quick access to diverse type of information. In the present competitive world, moving towards what we perceive as knowledge society, the access to right information at anytime, anywhere, about anything has gained high significance. This off course does not mean that the earlier societies were not aware of importance of information or were not knowledgeable. The information played very important role even in ancient time when people of the sub-continent evolved into agri-pastoral society, domesticated plants, animals and learned farming using draft animals, inventing tillage, seeding, intercultural operations, harvesting, and primary processing and prospered as interregional/ international traders. They were knowledgeable enough to evolve into present day society. The crucial difference now is the speed with which you can access information, the magnitude of available information and removal of geographical boundaries to access information. The developments in computer technology itself revolutionized the world and the sudden growth in telecommunication methodologies provided the necessary synergy to create a catastrophic change breaking every boundary and connecting the planet into one giant network of information and knowledge.

In such a scenario, the demand for authentic and credible digital information sources has risen in agriculture sector, be it research, education or extension. End to end value chain development requires quick access to diverse type of information. ICAR took several major digital initiatives to reform the way research and development is done in our traditional system in Agricultural Research and Education System (Consortium of e-Resources in Agriculture (CeRA); RKMP (Rice Knowledge Management Portal; http://www.rkmp.co.in/) Agroweb; Agripedia; etc.). CeRA is a platform for digital access to commercially subscribed research journals in consortium mode providing access to latest research all over the world directly from publisher websites through an aggregator [Chandrasekharan *et. al.*, 2012]. In addition to commercially available knowledge, the organizational knowledge available free within NARES is equally important for informed growth of research and education in agriculture sector in the country. Realizing this subproject E-Granth was initiated under National Agricultural Innovation Project (NAIP) under the aegis of Indian Council of Agricultural Research (ICAR) to enable digital access to vast amount of information available in the Indian National Agricultural Research & Education System (NARES), one of the largest agricultural research & education system in world (Jain *et. al.*, 2014, Jain *et. al.*, 2016 & NAIP, 2014). The objective of this project is to capture, collate and disseminate the organizational knowledge available in the form of research thesis, various technical reports, unpublished work, old valuable books and journals locked in libraries because of depleted condition, all kind of institutional publications and other knowledge resources. The details of architecture *viz.* storage, logical and application, methodology for mobile application and usage statistics of Krishikosh were explained in 2.1-2.4 respectively.

2. Methodology

2.1 Architecture of digital repository

Krishikosh platform is an Institutional Repository for collecting, preserving, and disseminating information in digital form for the intellectual output of an institution. It has been designed using open source software DSpace which has efficient Integrated Content Management System (ICMS), suitability configured to meet the requirements of NARES and created dependable digital repository. In this repository, some important terminology such as Community, Sub Community, Collection, Item and Bitstream needs to be understood. The flow chart for above mentioned terms is presented in figure 1. Community is the highest level in the hierarchy of DSpace logical model, which can be organizations / institutions / Universities. A community is organized into collections which consist of books, institutional publications, proceedings, reports, reprints, other publications and thesis for a specific community. The relationship between community and collection is many to many i.e. a community has many collection and vice-versa. The item is a single unit which means it consists of articles, thesis, reports, book etc. Each item has its respective Dublin Core metadata record. The Dublin Core is a schema which comprises of small set of vocabulary terms that can be used to describe the physical resources of item such as books, thesis, reports etc. Other metadata (copyright policy and non-exclusive distribution license) is stored in an item itself in the form of bitstream but the Dublin core metadata is further useful in interoperability and easy navigation of an item. The metadata that is information about the data of an item is first organized into bundles which consist of closely related information and then they form a bitstream. Every item has its bitstream. Each bitstream is then linked with one bit stream format for example if it is PDF/A or not. PDF/A format is international standard format not dependent on Adobe. The relationship between bitstream format and bitstream is one to many which means there can be only one bit stream format for a particular bitstream but many bitstream exists which has same bitstream format. These relationships are established to avoid duplication of content and easy management of system (Bass et. al., 2002). Kumar et. al. (2016) discussed the architecture of an open source DSpace especially for Krishikosh repository.

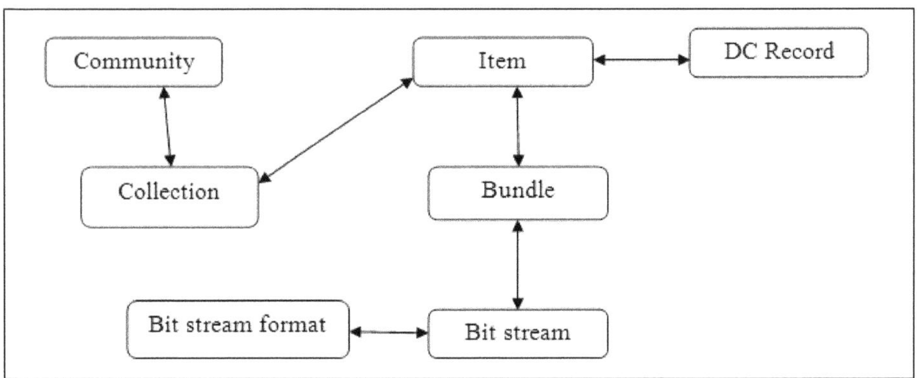

Figure 1. Logical DSpace data model

2.2 Architecture of DSpace

Dspace is based on three-layer architecture which consists of storage layer, logical layer and application layer and each layer includes different components as depicted in figure 2. The storage layer is responsible for physical storage of metadata and content. The logical layer deals with managing the content of the archive, users of the archive (e-people), authorization, and workflow. The application layer contains components that communicate with the world outside of the individual DSpace installation.

2.2.1 Storage layer

The lowest layer is the storage layer which consists of relational database, *viz.* postgres SQL (Structured Query Language) and bitstream storage module for storing the data and metadata respectively. The relational database management system (postgres SQL) is an open source licensed software i.e. there will be no barriers to implement Dspace anywhere or if the system deals with multiple instances (DSpace reference manual;2002). JDBC (Java database connectivity), which defines how a user may access postgres SQL. Bitstream storage systems are to provide storing the content by two different means *viz.* File system and Storage Resource Broker (SRB). The two units (postgres SQL and bitstream storage module) of the storage layer have Application Programming Interface (API) which is required to access the upper layer of the architecture.

2.2.2 Logic Layer

The architecture of logical layer consists of several modules such as core tools, administration toolkit, authorization, browsing tools, E-group/person, workflow, handle manager, content manager, history record etc. The layer deals with managing these contents. Each module has its own API through which the upper layer can be accessed. These APIs together is termed as Dspace public API. Core Classes consist of Configuration Manager, constants, context and log manager. Configuration Manager is responsible for reading the main dspace configuration properties file, managing the 'template' configuration files for other applications such as Apache, and for obtaining the text for e-mail messages. Constants are used to represent types of object and actions in the database. Any code that uses API in the logic layer must first create itself a Context object. Log Manager consists of a method that creates a standard log header, and returns it as a string suitable for logging. Content Management API, contains Java classes for reading and manipulating content stored in the DSpace system. The workflow system models the states of an Item in a state machine with various stages (submit, granting of license, archiving). The Submission workflow Manager is invoked by events as per Collection and its steps are defined by creating corresponding entries in the List named workflow group. Administration Toolkit contains classes for administering a DSpace system. To create Administrator class, a simple command-line tool was executed via /dspace/bin/createadministrator, that creates an administrator e-person with information entered from standard input. e-person/Group Manager class tracking the registered users in DSpace. The class has methods (set and get) to create an EPerson details such as first and last names, email, and password. There are find methods to find an EPerson by email (which is assumed to be unique,) or to find all e-people in the system. The authorization system in DSpace gave the 'policy state' for security purpose. The policies are attached to resources with the details who can perform the actions (READ, WRITE, ADD, etc.) through e-Person groups. The ADD and REMOVE actions must require this policy state. For creating an item ADD permission are granted for Collection, which contains Items. Separate policy checks for items and their bitstreams enables policies that allow publicly readable items, but parts of their content may be restricted to certain groups. Handle Manager/Handle Plugin creates and organizes components (plugins), and helps select a plugin in the cases where there are many possible choices. Handle Manager is used to create and look up Handles, and Handle Plugin is used to expose and resolve DSpace Handles for the outside world via the Corporation for National Research Initiatives (CNRI) Handle Server code. Handles are stored internally in the handle database table. Search API is the Lucene search engine which involves searching of task by indexing. It is the class, which contains index Content as an Item, Community, or Collection. Browse API maintains indices of dates, authors, titles and subjects, and allows users to extract parts of these items. The purpose of the history recorder is to capture a time-based record of significant changes in DSpace, in a manner suitable for later refactoring or repurposing.

2.2.3 Application layer

The uppermost layer is the application layer which communicates with the end user as it consists of web user interface. It also consists of web related services namely Statistical tools, OAI-PMH (Open Archives Initiative Protocol for Metadata Harvesting) data Provider, Simple importer/exporter, media filter and METS (Metadata Encoding and Transmission Standard) Exporter. The web user interface of DSpace is built on java servlet and JSP (Java Server Page) Technology, which allows user all over the world to access DSpace conveniently. OAI-PMH can be used to harvest bitstreams and metadata into DSpace from an external server. The layers only invoke the layer below it, for example the application layer can't invoke the storage layer directly. It helps in preparing digital resources and metadata by creating METS export. Media filters are used for transformation of file/bitstreams into a new content which may use for full text searching, it also creates thumbnails for an item that contained images.

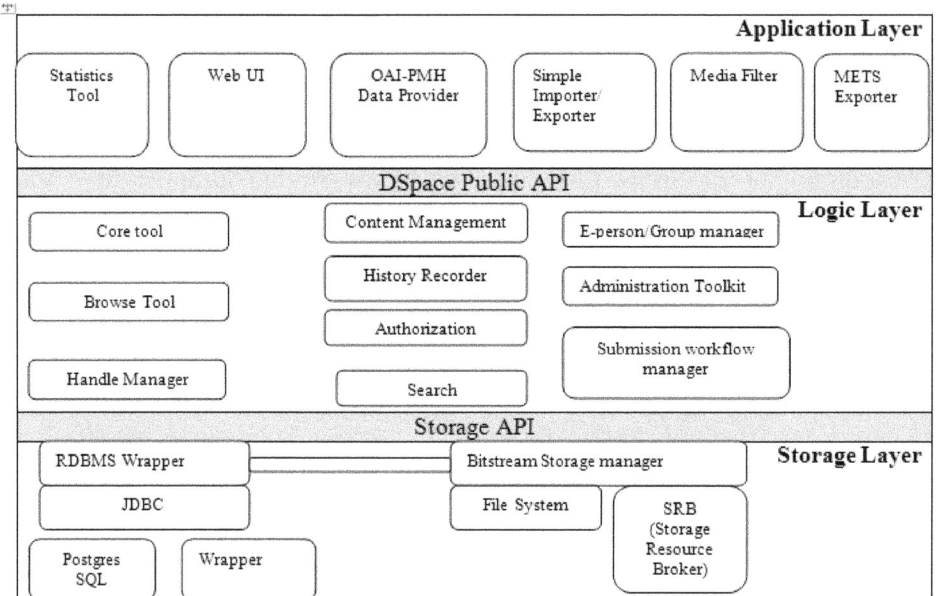

Figure 2. Architecture of DSpace

2.3 Mobile application for push notification

Krishikosh website is mobile responsive site which means that the design of the system responds or adapts depending upon the layout of the device (different mobile phones, tablets etc.). The responsive website improves the usability of the site. Making website responsive does not only means fitting the entire application on the users device but it also means to provide user a great experience while using the application and smartly pruning amount of information displayed. The responsive websites can be converted into mobile applications with the help of generating a web view in them. The Krishikosh application is developed in android studio. Java programming is used to generate codes for the web view of site. It is a fully functional web view in the application, user can register themselves on Krishikosh, can login and upload the documents, can view other reports and thesis anytime, anywhere from the application on their mobile. Krishikosh is a huge repository of thesis, reports, articles etc. Keeping track of all the information is very hectic and difficult for the user, which results in low usage of the Krishikosh site. Thus, an application with push notification was developed in order to engage the user with the Krishikosh, by sending messages regarding the new upload. It is implemented using GCM. For the testing purpose a sample database of the Krishikosh is taken and is implemented in PostgreSQl, it is the most advanced open source database. PHP script is used to connect the java programming in android studio, the database and the GCM server. The process of push notification is explained below:

1. First of all android device sends sender ID and application ID to the GCM server for registration.
2. When the devices are successfully registered with GCM, it will provide registration ID to android device.
3. After receiving registration ID, device will send this registration ID to the application server.
4. The application server will then store the registration ID into the database for later usage as and when required.
   a) Whenever new document is uploaded on the Krishikosh website, a push notification is needed, the application server is then sends a message to GCM server along with the device registration ID (stored in the database).
   b) GCM server will delivers the message to respected device using device registration ID.

The pictorial representation of push notification process is presented in Figure3.

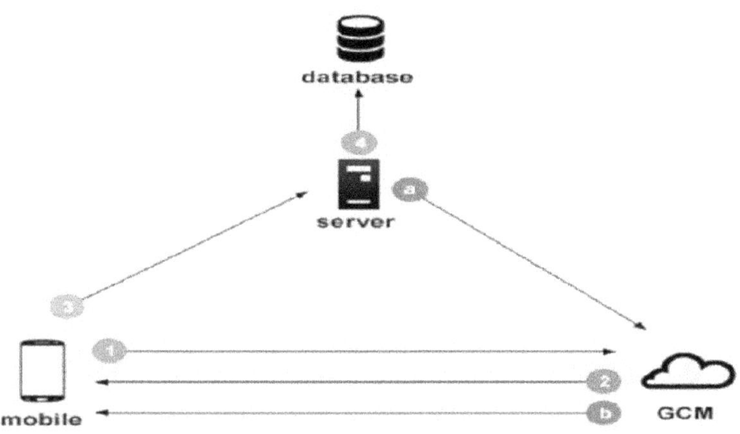

Figure 3. Architecture of Push Notification

Android Studio

For the development of Krishikosh android application, android studio version 1.5 was used. The system requirements of this version is operating system Windows/7/8 (32- or- 64- bit) but it can also be developed on Mac and Linux operating system., minimum 2 GB RAM, 2 GB of available disk space, Java Development Kit (JDK) 7, for accelerated emulator 64-bit operating system and Intel® processor with support for Intel® VT-x, Intel® EM64T (Intel® 64), and Execute Disable (XD) Bit functionality. After installing the android studio tool, required packages were installed using Android SDK (software development kit) manager as shown in Figure4. This also highlights the package Google APIs (x86 System Image) which is necessary to form the emulator device having Google APIs functionality

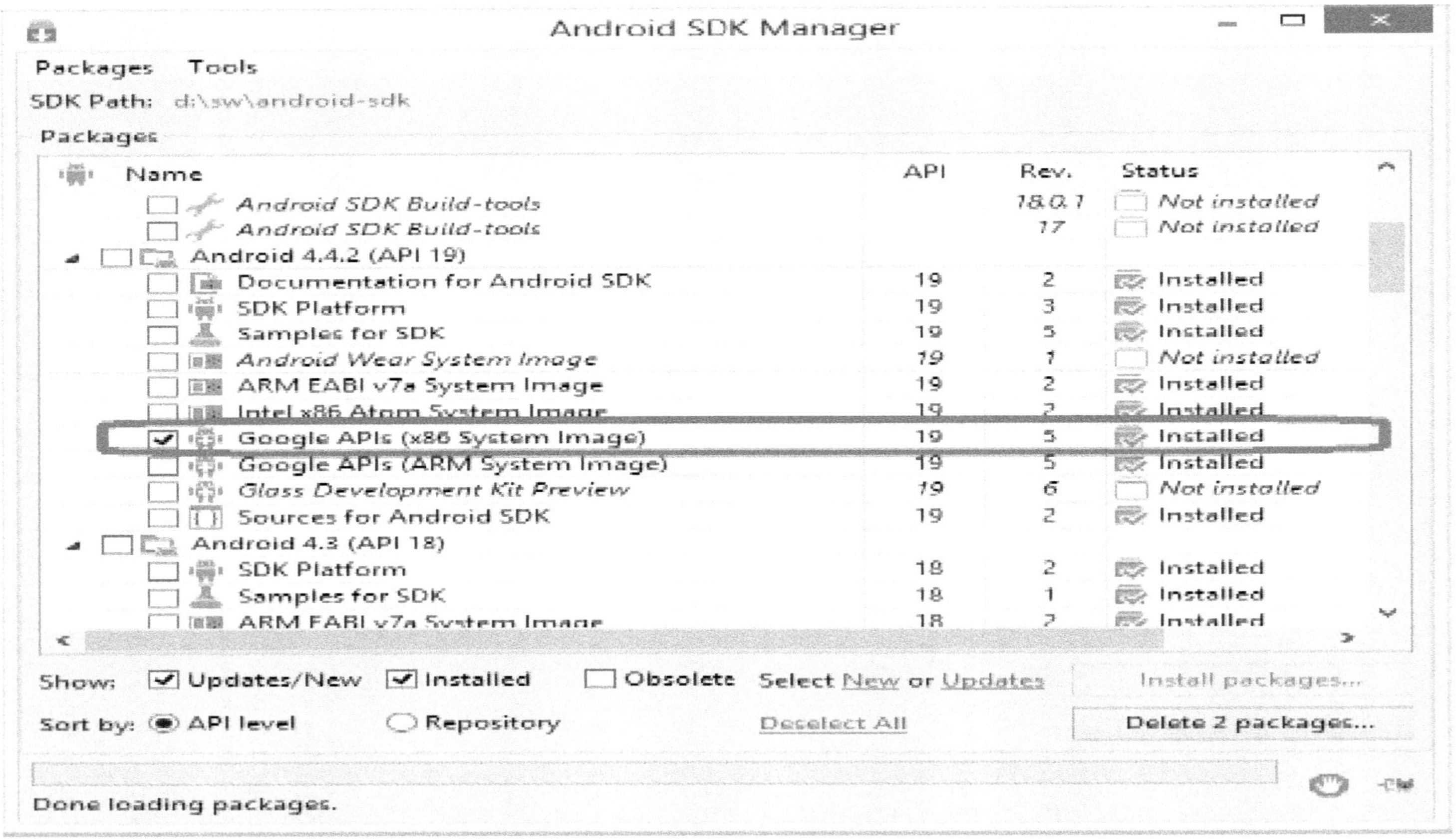

Figure 4.Packages installation in Android SDK Manager

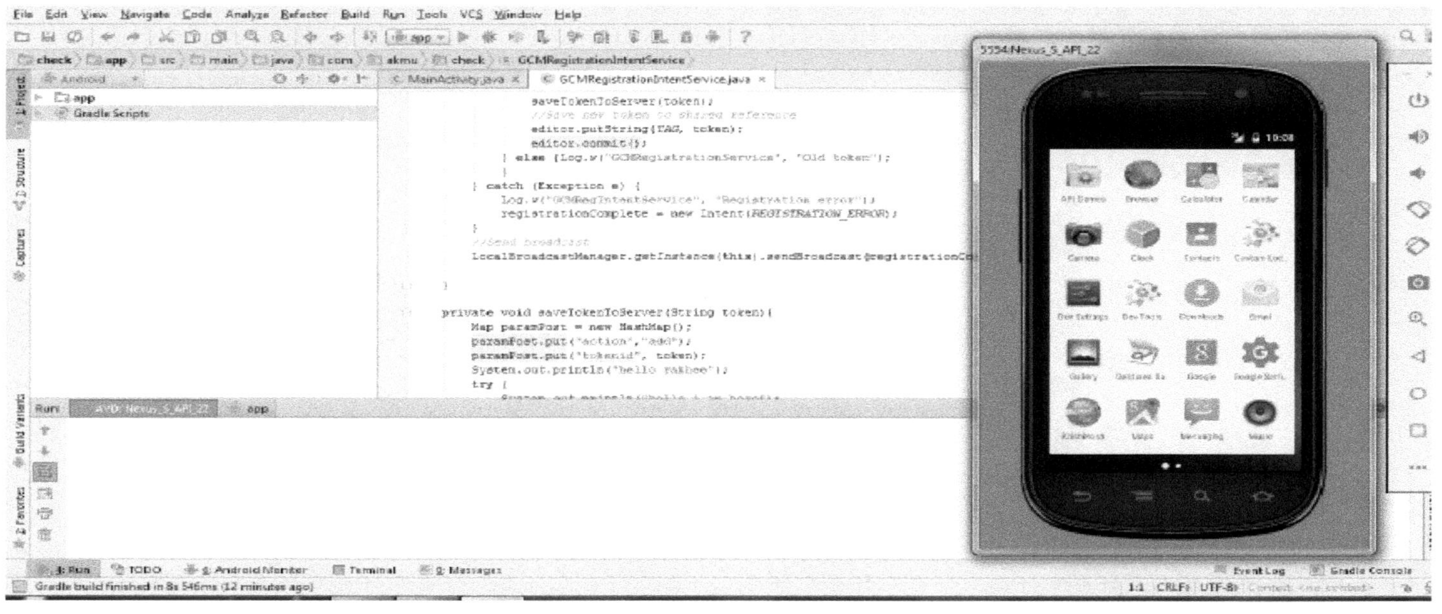

Figure 5. Emulator- A virtual device

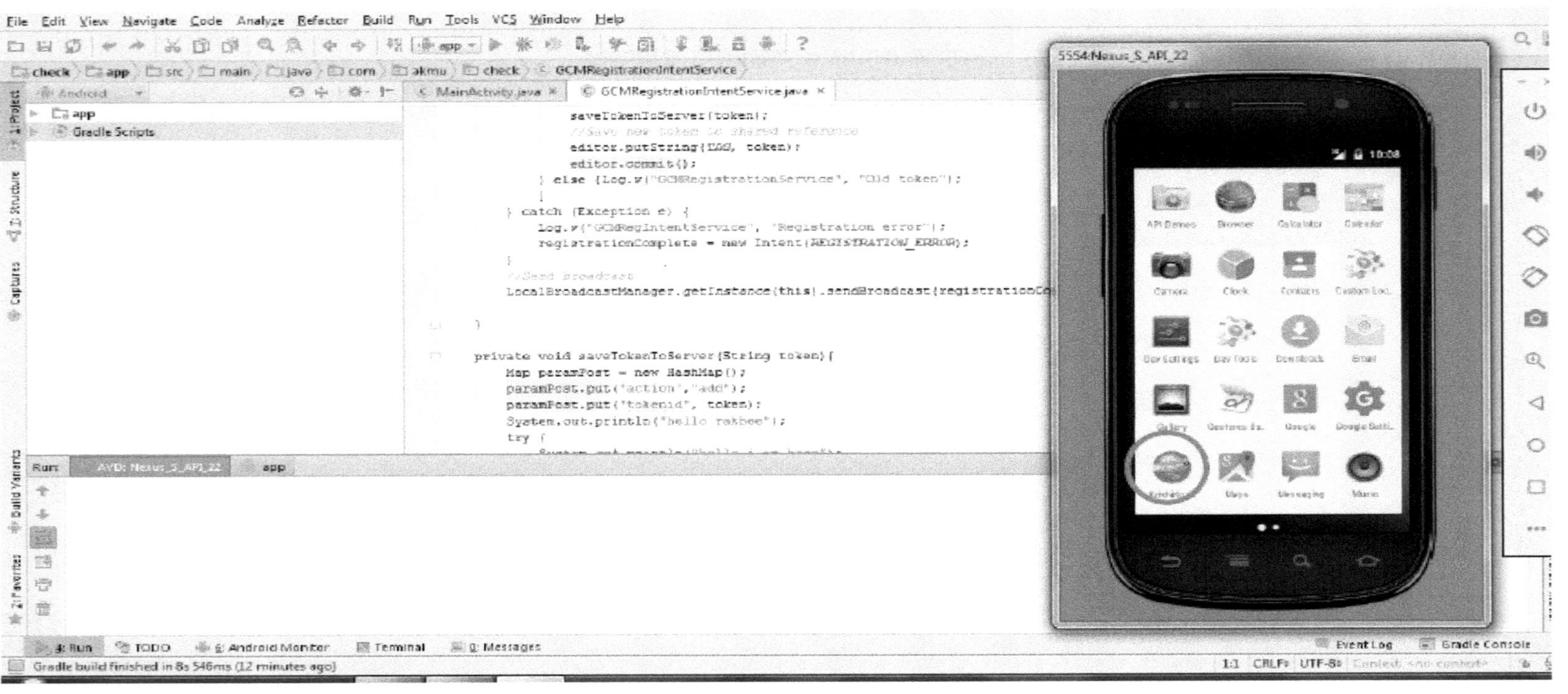

Figure 6. Krishikosh Icon in its mobile application

The emulator is a virtual device shown in Figure5, formed to test the application time to time. Instead of using an emulator, real device can also be used to test the application.

Android studio has most convenient way of drag and drop interface to make layout design. Code compilation on android studio is much better than other available software. Gradle integration in android studio is another great advantage to group all application files into one compressed file called Android Packages (APK). It is a Java Virtual Machine (JVM) based build system, developer can write their own script in Java, which Android Studio makes use of. Inside every project in android studio, there is build.gradle file, where developer can write their own script to automate the task. Android studio has native support for Google Cloud Platform. Google cloud platform allows running server side code using Google App Engine. It has many advantages; prime one is it can be used to create web-based application that can interact with Google Maps platforms. Further android studio also offers tools for integrating applications, running server side code locally for the testing purpose. Given below are the snapshots of mobile app in which the Krishikosh icon is generated (Figure 6), web view of Krishikosh in Mobile application (Figure 7), and thesis in readable format in Mobile application of Krishikosh (Figure 8).

2.4 Usages Statistics of Krishikosh

Usage statistics for community, collection, items are configured with Dspace to provide the usage details of the server to the administrators of Krishikosh (**http://wiki.duraspace.org**). Statistics on total visits of the communities, collections, items, etc.; countries along with cities from where the visits originate are available on this repository. The month wise details of bitstream views, item views, and searches performed, collection views, community views and user logins from June 2014 to August 2016 are utilized for analysis of data presented in Table 1.

3. Result and Discussion

3.2 Krishikosh Digital Repository of NARES

Krishikosh (http://Krishikosh.egranth.ac.in) is a versatile open access digital repository catering to the needs of NARES and has architecture of centralized hosting of content but decentralized management. The Krishikosh is hosted at the data center of ICAR-Indian Agricultural Research Institute (IARI), the premier research institute and deemed university under NARES. The hardware, software and connectivity is managed by Agriculture Knowledge Management Unit (AKMU) at ICAR-IARI and each institute or university can manage and administer it's own repository which is integral part of Krishikosh. Krishikosh is collectively managed, centrally aggregated repository with integrated search facility. Krishikosh is a digital Institutional Repository of important institutional publications including rare books and old journals, books, reprints, reports and thesis under open access policy of ICAR (http://icar.org.in/en/node/6609). The implementation of Krishikosh improves the accessibility coupled with preservation of institutional repository. This was achieved to create dependable digital storage and an efficient Integrated

Figure 7. Krishikosh web view in emulator

Figure 8. Thesis in readable format in mobile application

Content Management System (ICMS), an open source software DSpace. It provides following functionalities:

- **Improve Accessibility:-** The ICMS makes the holdings more accessible to scholars, teachers, academics and the general public, both within the premises as well as to those who cannot personally visit the NARES libraries but want to access the contents through the internet, under open access policy.
- **Enhanced Search ability:-** All holdings are grouped communities and collections based on institutions, subjects, themes or other criteria making large amount of information easily available on any subject matter for teaching, research and development. Any researcher looking for content on any subject or themes can have a unified access to content on all media types (manuscripts, photographs, audio-video, etc.) thereby making the searching much easier and faster.
- **Preservation:-** Preservation of all the rare documents in electronic form is an important objective. Also, once the documents are scanned and digitized, preservation of the originals can be ensured for a much longer period as the need to handle the physical documents is eliminated or minimized to a great extant since document are made available through the ICMS.
- **Content Selection:-** High power committees of subject matter experts have identified the content of intellectual and academic value to be included in the repository. Other institutions have identified the content in consultation with subject matter experts approved by the Directors/ Vice-Chancellors. The identified content was then harmonized centrally to avoid duplication.
- Various types of archival material at NARS comprises of rare books, old journals, reports, newsletters, annual reports, success stories, special bulletins, convocation addresses, endowment lectures, author's collections, preprints, reprints, patents, manuscripts, periodicals, grey literature, photographs, existing digital content, audio-video recordings.
- It is NARS's intention to make the Metadata for all records (and categories) freely available to all, however the actual records would be accessible based upon its access category.
- All of NARS's holdings are classified under the following three access categories:
- **Public Access** : Any record that can be made available to public at large shall fall under this category
- **Privileged Access:** Records classified under this category shall be accessible to only to those individuals or organizations that have a privileged status with NARS (such as other national / state archives / research and academic institutes / eminent researchers etc.). Others (the world at large) would have to seek prior permission / approval from IARI to access any Record classified as Privileged Access.

- **Prohibited Access:** Records which are accessible ONLY to NARS authorized officials, due to their confidential and sensitive nature as defined by statutory rules and regulation.

Thus, Krishikosh is a unique repository of knowledge in agriculture and allied sciences, having collection of old and valuable books, institutional publications, technical bulletins, project reports, lectures, preprints, reprints, thesis, records and various documents spread all over the country in different libraries of Research Institutions and State Agricultural Universities (SAUs). The home page of this repository is shown in figure9 and can be visited through the link (http://Krishikosh.egranth.ac.in/).

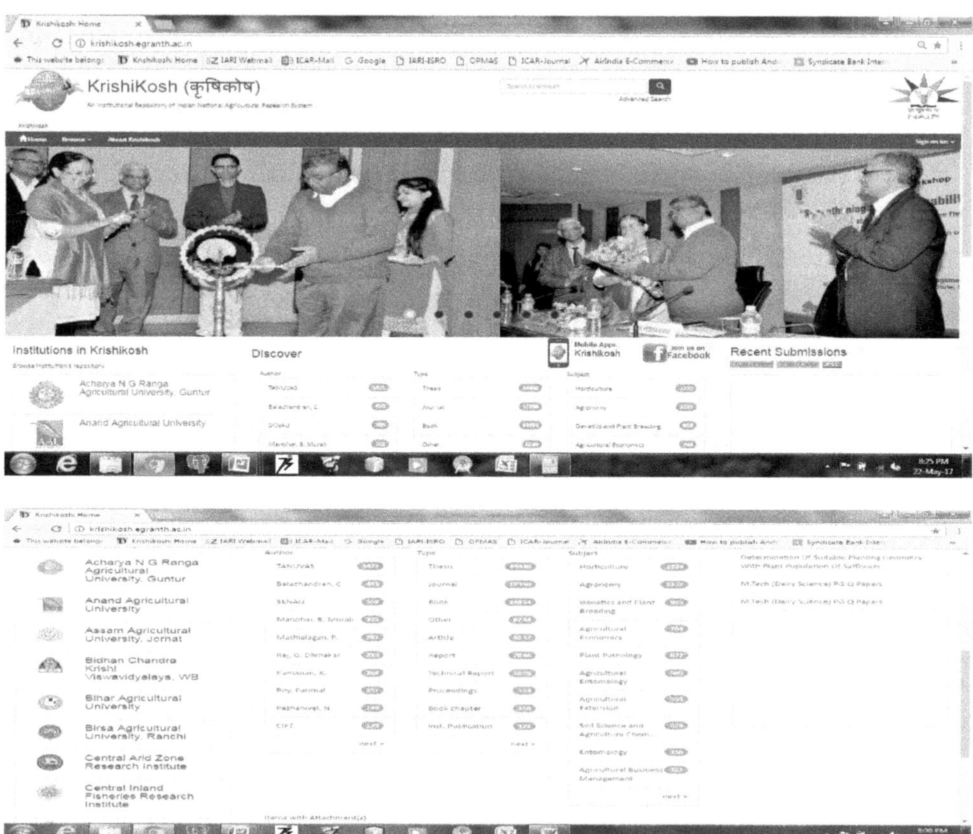

Figure 9. Home Page of Krishikosh

Salient Features of Krishikosh

- Krishikosh is digital repository platform capable of Decentralized Management of Content but Centralized Hosting and maintenance for convenience of multiple users.

- Each institution has its own repository with full control without effort of maintaining hardware/software which is centrally managed at IARI.
- Krishikosh Institutional repository of NARES provides open access to institutional knowledge.
- At present Krishikosh has more than 29 million digitized pages in 95,000 digital items (volumes) like old books, old Journals, reports, proceedings, reprint, research highlights, training manuals, historical records.
- Includes more than 47,000 theses plus other digitized contents for various centres.
- Agrotags developed at IIT Kanpur has been integrated for semantic search.
- Krishikosh uses PDF/A format (ISO19005-1:2005) which is international standard format not dependent on Adobe and suitable for archiving as it has all the dependencies build into the document for proper display even with the future technologies.
- Krishikosh provides ready software platform, similar to 'Cloud Service' for individual institution's self-managed repository with central integration of all the individuals repositories.
- Krishikosh is an open access Institutional Repository which has been developed by customizing open source software Dspace.
- It is full text searchable open access repository. Semantic search enabled through the integration of Agrotags.
- Comprehensive search and browse options are available.
- Easy to register for additional facilities.
- User feedback provision for continuous improvement.
- Subscribed users can get collection updates.
- Institutional users can administer their own repository and can upload, remove, set embargo on their institutional content.

Krishikosh platform is an Institutional Repository for collecting, preserving, and disseminating information in digital form for the intellectual output of an institution. In this repository, some important terminology such as Community, Sub Community, Collection, Item and Bitstream needs to be understood. The explanations of these terminologies are given below.

- **Community:** Community is the top level reference term which describes the University/ICAR Institute group. Generally the right to create a Community is with the Administrator of the Krishikosh.
- **Sub Community:** This is second level of hierarchy. It may describe departments/ division under the University/ICAR Institute.
- **Collection:** Collection is a part of Community or Sub-community in which we can add different categories like books, thesis, journals, newsletters etc. Creating collection is necessary to post the document under Krishikosh.

- **Item:** The record/document which is uploaded in collections is termed as item.
- **Bitstream :** It is the file which will be uploaded in the Krishikosh preferably a searchable pdf/a or pdf file.

3.3 Mobile Application

In Krishikosh repository large number of documents including the thesis was uploaded. It is very difficult to keep track of documents. Keeping track of all these information is very hectic and difficult for the user, therefore an android mobile application with push notification was developed for effective usage of Krishikosh repository in order to engage the user with the Krishikosh by sending messages regarding the new upload. The push notification is received using Google Cloud Messaging (GCM). It is a free service that enables developers to send messages between the servers and the client apps. This application is based on three tier architecture *viz.* presentation tier, application tier and data tier. Krishikosh mobile application was developed using android studio and coded in Java Programming with live database of Krishikosh using Postgres SQL. APK are Available at Google Playstore, IARI, Egranth and Krishikosh website In the later stage of development of mobile application user also have an advantage of categories the document based on community-wise, subject-wise etc.

3.4 Sorting of thesis uploaded on Krishikosh based on subject wise, theme wise and research problem wise

Initially, only Ph.D thesis are submitted in this repository, later on competent authority has decided to upload the M.Sc. thesis also (in meeting of 11th August, 2016 chaired by DDG(Agricultural Education). Krishikosh has the search facility based on Community (Institute/ SAUs), Collection (Books, thesis, Institutional publication, etc.,), item (name of author, year wise etc.). To provided the sorting or search facilities and their analysis based subject wise, theme wise and research problem wise, modified the form for uploading the documents for thesis by including the field Degree (M.Sc. and Ph.D), Subjects, Theme and Research problem basis in this repository. By adding these fields in the digital repository now we can sort the thesis / documents based on subject, degree, etc and further analysis can be performed. For searching the thesis based on above mention criterion, database is created which consist of previously submitted thesis with their respective degree, subject and their identifiers. This database will be linked with the current database of the Krishikosh repository. In the current database of the repository columns were created for degree, subject, their respective identifier and theme, research problem. Structured query language (SQL) is used to link the two databases and search the thesis in the current database for which there is no subject, degree and their respective identifier. Now the data from the sample database will be fetched and all the corresponding blank spaces will be filled. The updated database consists of the entire thesis (before and after adding the fields) with their subject, degree and all the other related information. This algorithm is represented in the form of flowchart in figure 10. The end-user of the Krishikosh repository may now be able to sort the thesis according to subject and degree

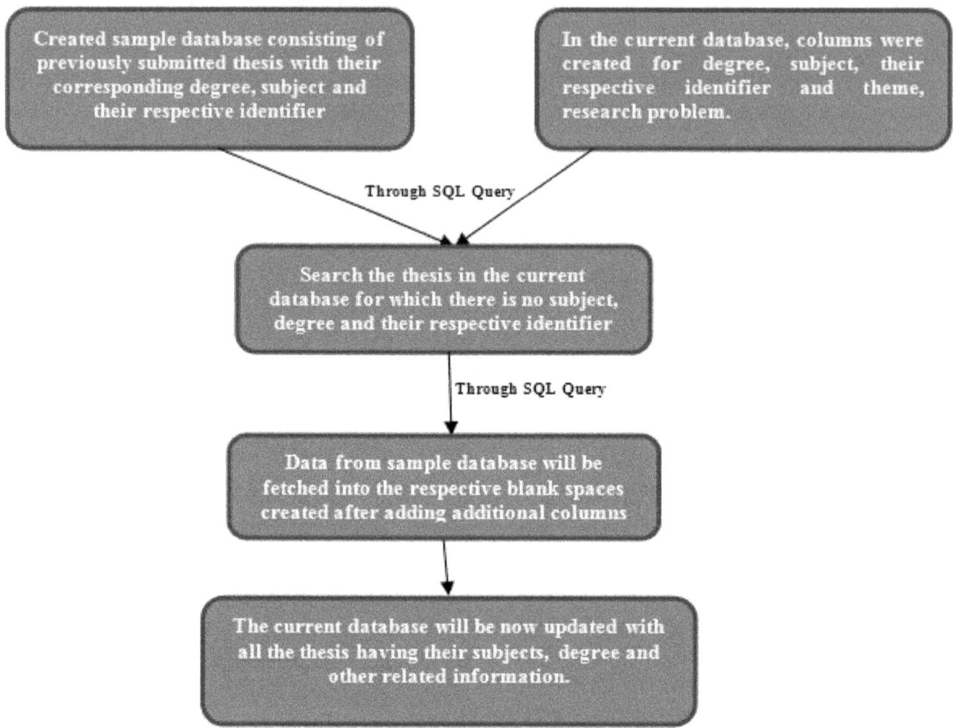

Figure 10. Flowchart for updating the databases with subject, degree, themes and research problem wise

3.5 Development of Copy Detection system in Krishikosh through keywords

Development of algorithm for duplication of works in thesis. Developed the process of Copy Detection system in Krishikosh through keywords. In the Krishikosh, keywords were generated through metadata of thesis (collection) using the agrotagger a software tool for automatic generation of keywords along with the thesis ids, subject, degree (M.Sc. /Ph.D.) etc. for selected community. Thesis id along with the keywords forms datasets which was extracted from the metadata of the thesis. These datasets were divided into two groups namely training and testing sets. Based on the training datasets, the database of the keywords (frequency based on occurrence of the words) was created through SAS software 9.3 available at our organization. Using pivot table maximum occurring keywords were identified. These identified keywords frequency was compared against the testing dataset of the thesis. If the occurrence of the keywords is more than the threshold (in our case study we have chosen the threshold value as 20, but it can vary based on the subject which can be decided by the subject matter expert). If the keywords of the test document is found to matched with the registered keywords more than the threshold (20 words in this case study), then the tested document may be considered as the case of duplicity, if the keywords are not matched with the registered keywords then there is no duplicasy in the document.

A case study

In the Krishikosh, keywords were generated through metadata of thesis (collection) using the agrotagger a software tool for automatic generation of keywords along with the thesis ids, subject, degree (M.Sc. /Ph.D.) etc. for selected community. Thesis id along with the keywords forms datasets which was extracted from the metadata of the thesis. These datasets were divided into two groups namely training and testing sets. Based on the training datasets, the database of the keywords (frequency based on occurrence of the words) was created through SAS software 9.3 available at our organization. Using pivot table maximum occurring keywords were identified. These identified keywords frequency was compared against the testing dataset of the thesis. If the occurrence of the keywords is more than the threshold (in our case study we have chosen the threshold value as 20, but it can vary based on the subject which can be decided by the subject matter expert). If the keywords of the test document is found to matched with the registered keywords more than the threshold (20 words in this case study), then the tested document may be considered as the case of duplicity, if the keywords are not matched with the registered keywords then there is no duplicasy in the document.

Metadata of thesis for National Dairy Research Institute (NDRI), Indian Agricultural Research Institute (IARI), Indian Veterinary Research Institute (IVRI), and TamilNadu veterinary and animal sciences university (TNVASU) were extracted from Krishikosh repository. These metadata consists of handler ID's, thesis title, subject, keywords, year, publishers etc. Keywords with respective Handler ID which is a unique identification for each thesis were extracted from the metadata. For each thesis there were about 10-15 keywords. These keywords are comma separated words, Using Excel's text to columns function, all the comma separated keywords were converted into columns format. Then SAS 9.2 code were developed to arrange the keywords in a single columns maintaining their respective handler ID's. The data is then sorted with the help of pivot table. The pivot table arranges all the keywords counting with their handler ID's keeping maximum appeared keywords in the top. Hence, the result is utilized to know which keywords appeared maximum number of times in which thesis. For creation of database of keywords, the 835, 467, 2295 and 1217 theses from NDRI, IVAR, IARI and TNVASU respectively were considered as training data. Based on these training sets the database of keywords / chuncks in domain of horticulture, animal sciences, dairy sciences, Genetics and plant breeding were developed. Rest keywords for other domain are under progress. These identified keywords were compared against the testing dataset of the thesis. If the occurrence of the keywords is more than the threshold (in our case study we have chosen the threshold value as 20, but it can vary based on the subject which can be decided by the subject matter expert). If the keywords of the test document is found to matched with the registered keywords more than the threshold (20 words in this case study), then the tested document may be considered as the case of duplicity , if the keywords are not matched with the registered keywords then there is no duplicasy in the document. For checking of duplicity in the thesis, the developed algorithm was tested as a case study in the subject of horticulture for a thesis available in the Krishikosh. For selected 300 thesis as test dataset from Krishikosh in the subject of horticulture. Document for the numbers of keywords that had similarity values less than 25 % were classified under No duplicity, between 25 to 60 % under Some what duplicity, between 60-80% under High duplicity and above 90% under Full duplicity. This criterion were on based intuitive methods, more objective techniques such as CART (Classification And Regression Tress), logistics, analytical Hierarchal (AHP) etc., may be used for this purpose. Based on the test data (keywords extraction form databases vs individual thesis) for 300 thesis under horticulture domain in Agriculture the summary statistics are presented in table 1.

Table1. Percentage of overlap of keywords

Category	% of overlap of Keywords
No duplicity	45.0
Some what duplicity	32.8
High duplicity	15.5
Full duplicity	6.7

3.6 Usages of Krishikosh

Usage statistics for community, collection, items are configured with Dspace to provide the usage details of the server to the administrators of Krishikosh (**http://wiki.duraspace.org**). Statistics on total visits of the communities, collections, items, etc.; countries along with cities from where the visits originate are available on this repository. The month wise details of bitstream views, item views, and searches performed, collection views, community views and user logins from June 2014 to August 2016 are presented in Table 2.

This table indicates that during the period from May to August there is rise in terms of searches performed and data view, this may be coincide with thesis submission time by most of the Universities.

The bitstream view, collection view and community view in 2016 (Jan-Aug) were presented in the figure 11-13. This indicates that all three views are maximum in the month from May to August in different years. This shows that the mostly review of literature were done in this month. This also coincides with the time of submission of thesis in most of the universities.

Table 2. Monthly hits / visit on Krishikosh for different collections, bitstream, user etc.

Year	Months	Bitstream Views	Item Views	Collection Views	Community Views	Searches Performed	User Logins
2014	**Jun**	372,578	92,487	16,754	12,327	16,307	388
	Jul	540,585	305,684	45,478	24,658	84,224	237
	Aug	261,066	115,304	12,674	9,556	79,724	20
	Sep	1,177,254	239,395	27,401	22,184	202,638	126
	Oct	414,100	76,458	10,820	8,328	103,460	37
2015	**Feb**	21,174	8,082	662	448	1,874	1
	Mar	376,978	85,512	8,322	4,784	110,951	51
	Mar	692,230	229,612	22,614	13,180	215,037	97
	Apr	231,746	60,708	5,460	3,145	31,675	20
	May	1,162,313	339,420	27,954	14,356	183,885	92
	Jun	968,623	277,740	31,412	37,696	153,825	90
	Jul	1,245,076	304,866	42,580	33,592	101,399	121
	Aug	1,055,054	322,674	28,068	16,358	129,709	135
	Sep	685,433	270,778	25,420	14,328	79,776	129
	Oct	680,962	271,162	30,088	23,238	59,755	132
	Nov	564,548	373,906	17,810	12,118	43,002	71
	Dec	624,714	213,376	22,842	18,990	54,328	125
2016	**Jan**	799,402	249,878	23,578	16,592	258,905	113
	Feb	866,056	294,948	26,026	19,960	394,464	362
	Mar	1,018,188	376,302	34,230	23,986	143,389	485
	Apr	855,717	296,174	30,494	25,950	119,550	258
	May	1,039,311	305,286	35,580	29,262	64,128	488
	Jun	1,063,607	400,406	41,332	31,418	78,333	729
	Jul	1,093,008	283,882	31,226	28,640	50,620	769
	Aug	640,829	124,050	17,190	15,594	25,742	579

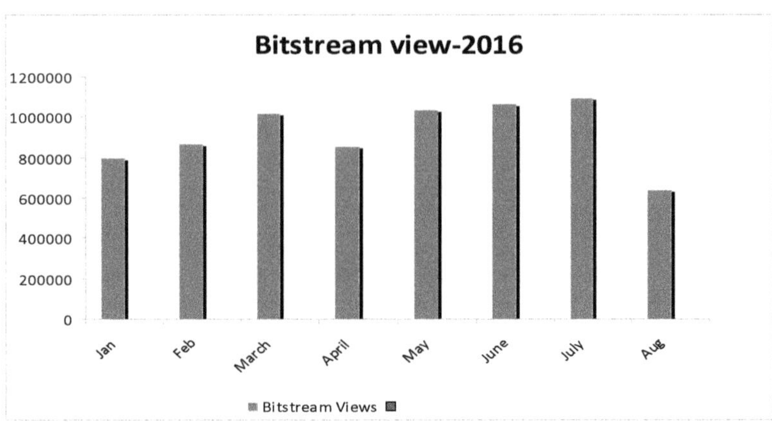

Figure 11

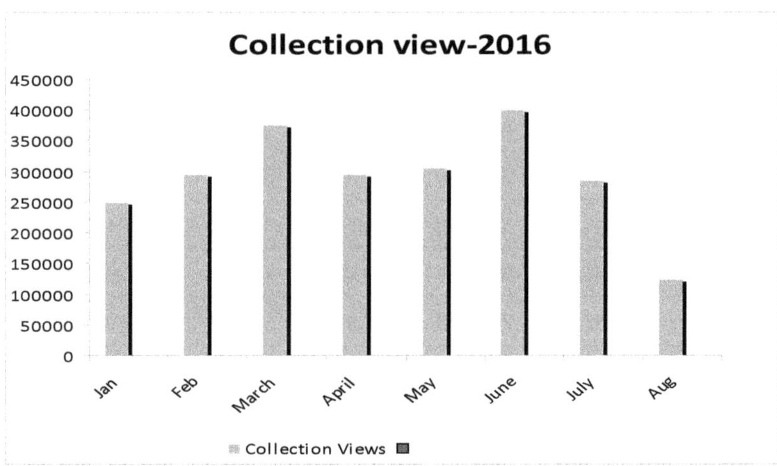

Figure 12

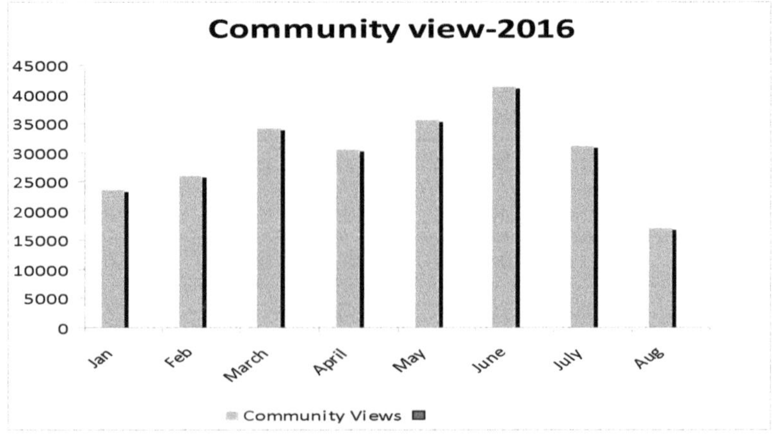

Figure 13

Krishikosh analytics during the period 1 April, 2016 to 31 March, 2017 represented in figure 14. It shows that there were about **1,27,558 users** who have been using the Krishikosh website. It also shows that during this time the new users were 70.8% of the current user. There were about **14,01,764** page views and 1.78 lakh session in this entire period. One session is equals to whatever user does on Krishikosh website (e.g. browses pages, reading, downloading Krishikosh app etc.) before they leave. India alone counts for 122,707 sessions followed by United States and China having 10,293 and 4,316 sessions respectively. The graph also represents the numbers of users vs. page views during this period. It is clear from the above graph that in January 2017 the page views were more. Krishikosh analytics also shows that the highest user (1,333) who has visited the Krishikosh website is on 20 January, 2017.

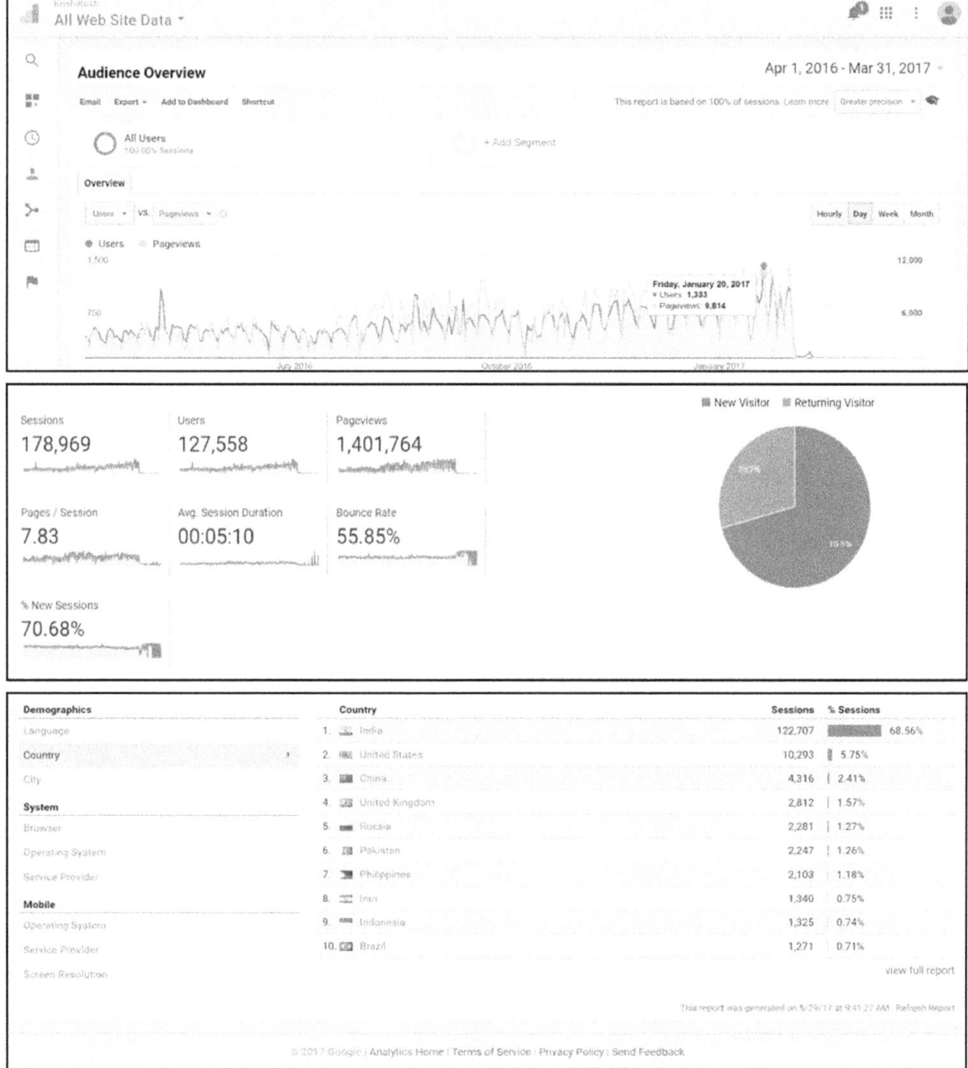

Figure. 14

At present Krishikosh has 20 million digitized pages in 98,000 digital items (volumes) like old books, old Journals, reports, proceedings, reprint, research highlights, training manuals, historical records, including 47000 theses digitized at various Universities in NARES (University / Institute wise thesis and other items in this digital platform are given in table 3). The months wise thesis submitted in Krishikosh are present in Fig. 15

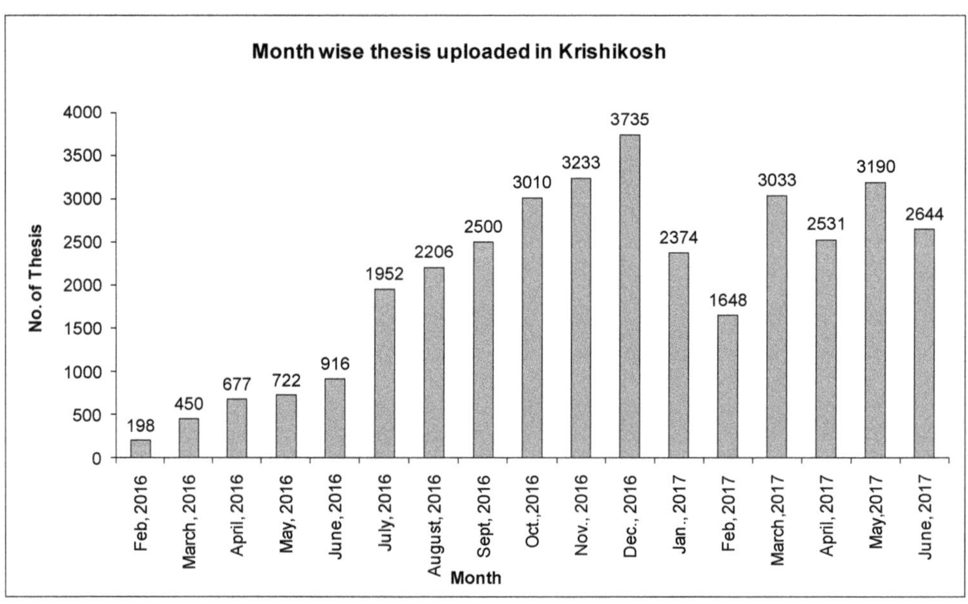

Figure15. Month wise submission of thesis in Krishikosh

Conclusion

ICAR has recently declared an Open Access policy. Krishikosh provides ready software platform to implement all aspects of the open access policy, similar to 'Cloud Service' for individual institution's self-managed repository with central integration of all individual repositories. At present Krishikosh has 16 million digitized pages in 71,000 digital items (volumes) like old books, old Journals, reports, proceedings, reprint, research highlights, training manuals, historical records, including 28500 theses digitized at various centers. The Krishikosh digital repository which provides digital platform for publishing can help and advice on IPR issues, research programme formulation and efficient management of institutional information assets. A customized digital repository platform for users of NARES Institutions, where they can upload and manage their own contents for compliance to open access policy of ICAR. Thus Krishikosh is a user-friendly platform to deposit, manage and access for Institutional information *viz.* books, reports, reprints, proceeding, thesis and Institutional publications produced by NARES researchers.

Table 3: University wise number of thesis and other items available in Krishikosh

S.No.	Name of the institution	Number of thesis	Number of other items	Total number items
1	Indian Agricultural Research Institute	4284	10979	15263
2	Tamil Nadu Veterinary and Animal Sciences University	2536	10693	13229
3	University of Agricultural Sciences, Bengaluru	2177	7942	10119
4	Professor Jayashankar Telangana State Agricultural University	6468	3167	9635
5	Indian Veterinary Research Institute, Izatnagar	0	4350	4350
6	Mahatma Phule Krishi Vidyapeeth, Rahuri	2452	1839	4291
7	National Dairy Research Institute	1950	1389	3339
8	Jawaharlal Nehru Krishi Vishwa Vidyalaya, Jabalpur	3185	8	3193
9	Chaudhary Charan Singh Haryana Agricultural University	2812	352	3164
10	Tamil Nadu Agricultural University, Coimbatore	1096	1397	2493
11	Indira Gandhi Krishi Vishwavidyalaya, Raipur	2185	139	2324
12	University of Agricultural Sciences, Dharwad	2109	0	2109
13	Indian Council of Agricultural Research	0	1946	1946
14	Anand Agricultural University	1124	352	1476
15	Punjab Agricultural University, Ludhiana	1429	4	1433
16	Indian Veterinary Research Institute-Mukteswar	0	1341	1341
17	Govind Ballabh Pant University Of Agriculture And Technology	1067	256	1323
18	Dr. Y. S. Parmar University of Horticulture & Forestry, Solan	1278	0	1278
19	Orissa University of Agriculture and Technology	744	155	899
20	Sri Karan Narendra Agriculture University, Jobner	423	432	855
21	Rajmata Vijyaraje Scindia Krishi Vishwa Vidyalaya, Gwalior	837	1	838
22	Junagadh Agricultural University, Junagadh	636	75	711
23	Maharashtra Animal and Fishery Sciences University, Nagpur	686	1	687
24	Navsari Agricultural University	561	8	569
25	Central Institute of Fisheries Education	271	278	549
26	Maharana Pratap University of Agriculture and Technology, Udaipur	512	0	512
27	Institute of Agricultural Sciences, Banaras Hindu University, Varanasi	0	455	455
28	Indian Institute of Oilseeds Research	0	422	422

(Continued)

Table 3: (Continued)

S.No.	Name of the institution	Number of thesis	Number of other items	Total number items
29	Karnataka Veterinary, Animal and Fisheries Sciences University, Bidar	397	22	419
30	National Bureau Of Agriculturally Important Insects	0	405	405
31	Indian Institute of Spices Research	0	390	390
32	Central Marine Fisheries Research Institute	0	348	348
33	Central Arid Zone Research Institute	0	319	319
34	Central Potato Research Institute	0	315	315
35	Sri Venkateswara Veterinary University, Tirupati	300	15	315
36	Dr. Balasaheb Sawant Konkan Krishi Vidyapeeth, Dapoli	283	7	290
37	Bidhan Chandra Krishi Viswavidyalaya, WB	279	0	279
38	Sher-e-Kashmir University of Agricultural Sciences and Technology, Kashmir	273	0	273
39	Central Inland Fisheries Research Institute	0	269	269
40	Dr.Y.S.R. Horticultural University, Venkataramannagudem	260	0	260
41	Rajasthan University of Veterinary and Animal Sciences, Bikaner	252	0	252
42	Chaudhary Sarwan Kumar Himachal Pradesh Agriculture University, Palampur	238	4	242
43	National Bureau Of Soil Survey And Land Use Planning	0	217	217
44	Bihar Agricultural University	58	157	215
45	Central Institute of Fisheries Technology	0	203	203
46	Kerala Agricultural University, Thrissur	198	0	198
47	CSA University of Agriculture and Technology	186	0	186
48	Dr. Panjabrao Deshmukh Krishi Vidyapeeth, Akola	167	0	167
49	Pandit Deen Dayal Upadhyaya Pashu Chikitsa Vigyan Vishwavidyalaya Evam Go-Anusandhan Sansthan, Mathura	158	0	158
50	National Institute of Research on Jute and Allied Fibre Technology	0	155	155
51	Central Soil Salinity Research Institute	0	153	153
52	Acharya N G Ranga Agricultural University, Guntur	137	0	137
53	PVNR Telangana Veterinary University, Hyderabad	131	0	131
54	National Bureau Of Plant Genetic Resources	0	128	128
55	Narendra Deva University of Agriculture & Technology	118	0	118
56	Birsa Agricultural University, Ranchi	112	0	112
57	Uttar Banga Krishi Viswavidyalaya, Cooch Behar, WB	111	0	111
58	Dr. Rajendra Prasad Central Agricultural University, Pusa	110	0	110

Table 3: (Continued)

S.No.	Name of the institution	Number of thesis	Number of other items	Total number items
59	Assam Agricultural University, Jorhat	104	0	104
60	Indian Institute of Water Management, Bhubaneswar	0	102	102
61	National Academy Of Agricultural Research Management	0	95	95
62	Indian Institute of Horticultural Research	0	91	91
63	Guru Angad Dev Veterinary and Animal Sciences University	89	0	89
64	Nanaji Deshmukh Veterinary Science University, Jabalpur	86	0	86
65	National Research Centre On Coldwater Fisheries - Bhimtal	0	85	85
66	Vasantrao Naik Marathwada Agricultural University, Parbhani	82	0	82
67	University of Horticultural Sciences, Bagalkot	60	20	80
68	Chhattisgarh Kamdhenu Vishwavidyalaya	70	0	70
69	Project Directorate on Poultry	0	68	68
70	Tamil Nadu Fisheries University, Thoothukudi	32	28	60
71	Rajasthan Agricultural University	54	0	54
72	Sam Higginbottom Institute of Agriculture, Technology and Sciences (SHIATS), Allahabad	51	0	51
73	Sher-e-Kashmir University of Agricultural Sciences and Technology-Jammu	48	0	48
74	West Bengal University of Animal & Fishery Sciences, Kolkata	48	0	48
75	Lala Lajpat Rai University of Veterinary & Animal Sciences	41	0	41
76	Central Institute of Freshwater Aquaculture	0	33	33
77	National Rice Research Institute	0	28	28
78	Indian Grassland and Fodder Research Institute	0	24	24
79	Uttarakhand University of Horticulture and Forestry, Bharsar	19	0	19
80	University of Agricultural & Horticultural Sciences, Shivamogga	3	6	9
81	Kerala University of Fisheries and Ocean studies, Ernakulam	5	0	5
82	Indian Agricultural Statistics Research Institute	0	3	3
83	Central Institute for Subtropical Horticulture, Lucknow	0	2	2
84	University of Agricultural Sciences,Raichur	2	0	2
85	National Research Centre on Camel, Bikaner	0	1	1
86	Total	45384		97058

References

Bass, M. J., Stuve, D., Tansley, R., Branschofsky, M., Breton, P., Carmichael, P., & Ng, J. (2002). DSpace–A sustainable solution for institutional digital asset services–spanning the information asset value chain: ingest, manage, preserve, disseminate. Internal reference specification: Technology and Architecture. Retrieved from http://web.mit.edu/dspace/live/implementation/design_documents/archite cture.pdf (accessed June 8, 2014).

Chandrasekharan, H., Patle, Sarita, Pandey, P S, Mishra, A K, Jain, A K, Goyal, Shikha, Pandey, Amit, Khemchandani, Usha and Kasrija, Rajkumari (2012). CeRA - the e- journal Consortium for National Agricultural Research System; *Current Science*, 102 (06): 847-851

DSpace –Internal Reference Specification (2002), Technology and Architecture.

DSpace reference manual (2002), The DSpace Foundation. [http://www.dspace.org/].

http://wiki.duraspace.org.

http://www.rkmp.co.in/

Jain, A. K., Chandrasekharan, H. and Kumar Rajesh (2014). Final Report of NAIP sub-project 'Digital Library and Information Management under NARS (e-GRANTH)'. Indian Agricultural Research Institute, New Delhi, India

Jain, A. K., Kumar, Amrender, Batra, Kamal, Kapur, Sanjiv (2016). Reference Manual on Krishikosh - A Repository for NARES. ICAR– Indian Agriculture Research Institute, New Delhi, pp50

Kumar A, Sharma R, Sharma A, Batra k, Kapur S (2016). Digital Initiatives for Agricultural Research & Education under ICAR in India. pp49- 68.

NAIP (National Agricultural Innovation Project) (2014). An Initiative Towards Innovative Agriculture. Final Report, National Agricultural Innovation Project, ICAR, New Delhi

Rathinasabapathy, G.; Veeranjaneyulu, K. and Amarendar Kumar (2016). Krishikosh: Institutional Repository of National Agricultural Education and Research System (NARES) of India: An Analytical Study. ICDL 2016 - Smart Future: Knowledge Trends that will change the world, New Delhi. pp. 873-887. ISBN: 9788179936535

5

Persistent Researcher and Author Identification Systems in the Networked Digital Environment

Dr. G.Rathinasabapathy,Ph.D.,

University Librarian
Tamil Nadu Veterinary and Animal Sciences University
Chennai- 600 051, Tamil Nadu
e-mail: librarian@tanuvas.org.in

ABSTRACT

The name ambiguity is one of the major problems faced by academicians and researchers across the globe. To circumvent the author name ambiguity problem, the need for assigning each researcher a 'unique author identifier' was realised. Though the idea of a centrally administered system to unambiguously identify authors of scientific papers has been around since the 1940s, it has received renewed attention with the proliferation of online journals, databases and open access publication archives in the networked digital environment. Unique identifiers provide a number of benefits to researchers, authors, institutions, publishers, funding organisations and scholarly societies. Many initiatives were taken to identify the authors with persistent identity viz., AuthorClaim (1999), LATTES (1999), VIAF (2003), NARCIS (2004), arXiv Author

ID (2005), Scopus Author Identifier (2006), Names Project (2007), ResearcherID (2008), ORCID (2010) and ISNI (2012). This paper attempts to profile the important author identification systems.

Keywords: *Researcher Identification system, Author Identification system, AuthorClaim, LATTES, VIAF, NARCIS, Names Project, arXiv, Scopus Author Identifier, ResearcherID, ORCID, ISNI*

1. Introduction

In the scholarly communication we can see many authors with similar names. Because multiple researchers in the same or different fields may have the same first and last names, there is an author ambiguity problem within the scholarly research community and it is a global problem.

Problems arise in numerous ways *viz.*, (i) several researchers having the same or similar names, researchers sometimes publishing under name variations (ii) researchers moving from institution to institution (iii) the increased frequency of researchers crossing strict disciplinary lines to work in other areas or collaborate with those in other areas.

For example, in India, if an author name is mentioned as A.K. Jain, it may be Arun Kumar Jain, Ananda Krishan Jain, Abhay Kumar Jain, Akash Kumar Jain, Abhilash Kumar Jain, *etc.*

In some cases, the author name is published in different ways. For example the name of an author is G.Dhinakar Raj. His name is published by the journals in four different types e.g. (i) GDR Raj; (ii) G.Dhinakar Raj; (iii) Gopal Dhinakar Raj; and (iv) G.D.Raj. Thus, the name ambiguity is one of the major problems faced by the academicians and researchers across the globe.

2. Author Identification System

To circumvent the author name ambiguity problem, the idea has been raised of assigning each researcher a 'unique author identifier'. The idea of a centrally administered system to unambiguously identify authors of scientific papers has been around since the 1940s, but has received renewed attention with the proliferation of online journals, databases and open access publication archives in the networked digital environment. To overcome the author name ambiguity, need for unique author identification was felt by the academic and research community.

3. Types of Author Identification Systems

There are two types of author identification systems available while few systems are national level and few other systems are global.

Many initiatives were taken to identify the authors with persistent identity *viz.*, LATTES (1999), VIAF (2003), NARCIS (2004), arXiv Author ID (2005), Scopus Author Identifier (2006), Names Project (2007), ResearcherID (2008), ORCID (2010) and ISNI (2012).

Among this the LATTES, NARCIS, NamesProject are national level author identification systems of Brazil, Netherlands and U.K. respectively. The remaining are global level author identification systems. A list of author identification systems are furnished in the following table.

Name	Organisation	Kind	Disciplines	Countries	Year started
AuthorClaim	Open Library Society	Non-profit	All (mostly economics)	All	1999
LATTES	National Council for Scientific and Technological Development	Government	All	Brazil	1999
VIAF	Online Computer Library Centre (OCLC) and 15 national libraries	Non-profit	All	Several	2003
NARCIS	Royal Netherlands Academy of Arts and Sciences	Government	All	Netherlands	2004
ArXiv Author ID	Cornell University Library	Academic	Physics, maths, computer science and related disciplines	All	2005
Scopus AuthorID	Elsevier	Commercial	All	All	2006
Names Project	Mimas, British Library	Academic	All	UK	2007
ResearcherID	Thomson Reuters	Commercial	All	All	2008
ORCID	ORCID	Non-Profit	All	All	2009
PubMed AuthorID	National Library of Medicine (NLM)	Government	Life Sciences	All	2010
ISNI	International Organization for Standardization	Non-Profit	All	All	2012

3.1 AuthorClaim

The AuthorClaim registration service aims to link scholars with the records about the works that they have written, as recorded in a bibliographic database. The authorclaim

- builds a profile for the author, that shows all his/her identified works.
- Users of bibliographic databases that use AuthorClaim record can link right to his/her profile page or their homepage.
- It becomes possible to distinguish the works of homonyms, a people that are recorded with abbreviated names. Thus if you are recorded in the bibliography as "J. Smith", it becomes possible to see that you are "Jane Smith" rather than "James Smith".

- As and when such services are in place, the author can get regular statistics about downloads and citations of your works.
- The collected data can be used to compute various rankings.

Any author can create a profile and an account in the system. The Registration process requires a valid email address and takes six screens to go through. These screens include a search for the author's workplaces, as well as a search for his/her publications. Then the system sends a confirmation email to the author. When the author opens the confirmation link in the email message, his/her data are saved and account activated. The service is in a building-up phase. It is not experimental. That means, author records that users create now will be kept in perpetuity.

3.2 LATTES

The Lattes Platform started in 1999 is an information system maintained by the Brazilian Government to manage information on science, technology, and innovation related to individual researchers and institutions working in Brazil. It is named after a Brazilian physicist, Cesar Lattes, and it is maintained by the federal bureau responsible for funding science and technology at the federal level National Counsel of Scientific and Technological Development.

Since all researchers and institutions are required to maintain their records up to date, the Lattes Platform can be used not only to obtain information on individual researchers but also to conduct performance evaluations at the organizational level.

3.3 Virtual International Authority File (VIAF)

Virtual International Authority File, is an international cooperative effort amongst nearly two dozen national (and smaller) libraries to facilitate unambiguous identification through linking of author and individual authority records across major bibliographic databases and languages, worldwide.

3.4 National Academic Research and Collaborations Information System (NARCIS)

NARCIS has been developed by the KNAW to increase visibility and retrievability of Dutch scientific research. This development takes place in close cooperation with the Dutch universities, NWO (Netherlands Organisation for Scientific Research) and other research institutes. NARCIS gives access to scientific information consisting of (open access) publications from the repositories of all the Dutch universities, KNAW, NWO, and a number of research institutes, the datasets of the institute DANS, as well as descriptions of research projects, institutes and researchers. This means that NARCIS cannot be used as an entry point to access complete overviews of publications of researchers (yet). On a national scale there are plans, however, to incorporate the publication data from the academic Metis-systems into NARCIS. By doing so, it will become possible to create much more complete publication lists of researchers.

3.5 arXiv Author ID

arXiv is an open access repository which gives access to more than 1,281,000 e-prints in Physics, Mathematics, Computer Science, Quantitative Biology, Quantitative Finance and Statistics by Cornell University Library, started in 2005.

It is a long-term goal of arXiv to accurately identify and disambiguate all authors of all articles in arXiv. Since 2005 arXiv has used authority records that associate user accounts with articles authored by that user. The use of public author identifiers as a way to build services upon this data is new in 2009. Initially, users must opt-in to have a public author identifier and to expose the record of their articles on arXiv for use in other services. arXiv can also be linked to ORCID.

It would also be beneficial to associate author records in arXiv with author records in other scholarly communication system, for example with the INSPIRE database in high-energy physics. Association of author records across different systems would facilitate the creation of services and tools that operate over multiple repositories, or combine data from multiple sources.

3.6 Scopus Author Identifier

Scopus is the world's largest abstract and citation database of peer-reviewed research literature covering 22,000 titles from more than 5,000 international publishers. Scopus Author ID is a product from Elsevier which was started in 2006. It is a commercial service and it covers all disciplines and all countries.

The Scopus Author Identifier distinguishes between these names by assigning each author in Scopus a unique number and grouping together all of the documents written by that author.

This feature is especially useful for distinguishing between authors who share very common names like Smith or Wang or Lee. Additionally, author names in Scopus can be formatted differently. For example, the same author could appear in one document as Lewis, M; in another as Lewis, M.J; and in another as Lewis, Michael. Scopus Author Identifier matches the documents of this author and groups these name variants together so that authors, even if cited differently, are identified with their specific papers. This helps you find and recognize an author, despite variations in name spelling.

Author will be automatically assigned a Scopus Author ID when the author publishes in a journal indexed by Scopus. The author need not to register for a Scopus Author ID, if the author has a paper indexed in their database, the author is automatically assigned a Scopus Author ID.

Having a Scopus Author ID gives the author the following benefits:

- Automatic author identification
- Publication list
- Citation metrics available

3.7 Names Project

Names project was funded by the Joint Information Systems Committee (JISC) and began in July 2007. It was funded to investigate requirements for a name authority service for UK repositories. A prototype name authority system has been developed as part of this work and a number of connections have been made with UK stakeholders and with international projects working in a similar space.

Working in partnership with the British Library, and using data from our Zetoc service, the system currently identifies over 50,000 of the UK's top researchers. This project works with universities and other research institutions to increase the number of individuals who have a Names identifier.

3.8 ResearcherID

ResearcherID is a website where invited researchers can register for a unique ResearcherID number. At this site, users can: Update their profile information; Build their publication list using Web of Science search services or uploading a file; Select to make their profile public or private.

Registered as well as non-registered users can search the Researcher Registry to view profiles and find potential collaborators. A ResearcherID number is a unique identifier that consists of alphanumeric characters. Each number contains the year it is registered.

We can search the registry by entering the ResearcherID number you are interested in, or by searching by last name and first initial combination, keywords, institution, or country. Alternatively, there are several visual search features that we can use to launch a search:

- Top Keywords: Displays the 100 most frequently occurring keywords
- Top Countries: Displays the 100 most represented countries (based on primary institution)
- Map: Displays membership distribution on a world map.

Currently, there are two ways to register for a ResearcherID account. Web of Science subscribers can register from the Web of Science homepage. Otherwise, a request may be sent for registration. ResearcherID is now integrated with EndNote. The researcher will be able to access his/her ResearcherID publication list in EndNote and take advantage of EndNote features for outputting their publications.

3.9 Open Researcher & Contributor ID (ORCID)

The Open Researcher & Contributor ID (ORCID) is the latest initiative, which is meant to absorb all positive elements of previous author and researcher ID schemes. Its identifiers are presented as Uniform Record Locators (URLs) or in a short form of 16 characters.

ORCID was officially launched in October 2012, and the number of registrants reached 3,603,217 by June, 2017.

The ORCID registry presents a unique opportunity to solve the problem of author name ambiguity. At its core the value of the ORCID registry is that it crosses disciplines, organizations, and countries, linking ORCID with both existing identifier schemes as well as publications and other research activities.

By supporting linkages across multiple datasets – clinical trials, publications, patents, datasets – such a registry becomes a switchboard for researchers and publishers alike in managing the dissemination of research findings. We describe use cases for embedding ORCID identifiers in manuscript submission workflows, prior work searches, manuscript citations, and repository deposition.

Eg. http://orcid.org/0000-0001-5109-3700, or 0000-0001-5109-3700

The long list of ORCID supporters now includes the world's leading universities, the British Library, large publishers such as Elsevier, Springer, Nature Publishing Group and Dove Press, and major funders such as the Welcome Trust and the National Institutes of Health.

The ORCID scheme supports multiple languages, and therefore, holds promise for increasing international visibility of researchers and authors working in non-Anglophone countries. Spanish, French and Chinese interfaces of ORCID are now available, while Korean, Japanese and Russian character sets can be added in the foreseeable future. The ORCID compatibility with open repositories, digital libraries and platforms such as CrossRef, PubMed Central, ScienceCentral, and KoreaMed Synapse made its IDs particularly useful for fast and transparent transfer of scholarly information globally.

With the ORCID initiative now becoming global and aiming at comprehensive recording of information about researchers and authors, it is envisaged that some of the persistent problems with publication ethics, such as inappropriate scientific authorship and nondisclosure of conflicts of interests, will be also curbed.

ORCID IDs can be integrated with information about all scholarly activities of the registrants, which can increase the transparency and decrease the rate of guest authorship in multi-authored papers. Knowing that their background and relationships with competing organizations is under scrutiny, authors, reviewers and editors will take an extra effort to properly declare secondary interests or refrain from publishing conflicting data. Publishers, in turn, can help their contributors, and primarily corresponding authors, responsible editors and reviewers by advising them to register with ORCID and to keep their accounts updated. And revising journal instructions by adding a relevant point on researcher and author identifiers can perhaps be the best option.

3.10 International Standard Name Identifier (ISNI)

The idea of unique identifiers materialized in 2012 by the development of the International Standard Name Identifier (ISNI) under the auspices of the International Organization for Standardization.

ISNI is the ISO certified global standard number of identifying the millions of contributors to creative works and those active in their distribution, including researchers, inventors, writers, artists, visual creators, performers, producers, publishers, aggregators and many more. It is part of a family of international standard identifiers that includes identifiers of works, recordings, products and right holders in all repertoires, e.g. DOI, ISAN, ISBN, ISSN, ISRC, ISTC, ISWC, *etc*.

The mission of the ISNI International Authority (ISNI-IA) is to assign to the public name(s) of a researcher, inventor, writer, artist, performer, publisher, etc. a persistent unique identifying number in order to resolve the problem of name ambiguity in search and discovery; and diffuse each assigned ISNI across all repertoires in the global supply chain so that every published work can be unambiguously attributed to its creator wherever that work is described.

ISNI holds public records of over 9.34 million identities, including 8.739 million individuals of which 2.599 million are researchers and 606,872 organisations.

The ISNI database is a cross-domain resource with direct contributions from 40 sources, including the Virtual International Authority File (VIAF), an aggregation of data from major national and research libraries.

3.11 PubMed Author ID

The National Library of Medicine, National Center for Biotechnology Information (NCBI) is developing a system that will address the problem of ambiguous author names within PubMed and facilitate accurate search and retrieval of a participating author's works. The specifics of PubMed Author ID, as the system is now known, are still evolving. It is currently envisioned that authors (or their designees) would register for the service through My NCBI and identify their research articles in PubMed using provided tools; this identification of articles will allow NCBI to link alternate names/spellings associated with an individual. The anticipated launch for PubMed Author ID is in mid-2011.

NLM has already laid the foundation for the system by developing a process for NIH-funded authors to identify their articles for grant reporting purposes. NLM expects to make PubMed Author ID interoperable with multiple external author ID systems, such as those developed by publisher groups, non-profit organizations, and other countries. NLM has not yet identified external author ID systems that it will incorporate in PubMed Author ID, but will work with outside groups as systems are developed in this rapidly evolving area.

Currently, NLM is not developing a PubMed Author ID system. However, NLM accepts author identifiers from outside organizations like ORCID when supplied to us by publishers with citation data. Also, PubMed disambiguates authors using the Computed Author display sort.

4. Conclusion

Author identification is a complex problem and involves a large number of stakeholders who sometimes have opposing views on some of the issues that need to be addressed. Building an author identifier system is therefore not just about technical challenges, it also requires decisions about openness, privacy, collaboration, business models and other critical issues. Though there are many such unique author identification systems, the Scopus AuthorID, ResearcherID of Thomson Reuters and ORCID are very familiar and growing steadily. Since the author identification is very important, the Library and information professionals should disseminate the significance of the systems to the researchers and academicians and encourage them to register.

References

Aerts R. Digital identifiers work for articles, so why not for authors? *Nature*. 2008 Jun; 453(7198):979. DOI: http://dx.doi.org/10.1038/453979b

Cals JW, Kotz D. Researcher identification: the right needle in the haystack. *The Lancet*. 2008 Jul; 371(9631):2152–2153. DOI: 10.1016/S0140-6736(08)60931-9

Thorisson GA. Accreditation and attribution in data sharing. *Nature Biotechnology*. 2009 Jan; 27(11):984–985. DOI: 10.1038/nbt1109-984b

Enserink M. Scientific publishing. Are you ready to become a number? *Science*. 2009 Mar; 323(5922):1662-4. DOI: 10.1126/science.323.5922.1662

Researcher Identifcation Primer. *GEN2PHEN Knowledge Center*. 2009. Available from: http://www.gen2phen.org/researcher-identification-primer

Habibzadeh F, Yadollahie M. The problem of "Who". *The International Information & Library Review*. 2009 Jun ;41(2):61-62. DOI: 10.1016/j.iilr.2009.02.001

Credit where credit is due: The Open Researcher and Contributor ID (ORCID). *Nature*. 2009; 462(7275):825. DOI: 10.1038/462825a

ORCID or how to build a unique identifier for scientists in 10 easy steps. *Gobbledygook Blog*. 2010. Available from: http://blogs.plos.org/mfenner

Warner S. Author Identifiers in Scholarly Repositories. *ArXiv*. 2010. Available from: http://arxiv.org/abs/1003.1345

Lane J. Let's make science metrics more scientific. *Nature*. 2010 Mar;464(7288):488–9. DOI: 10.1038/464488a

Qiu J. Scientific publishing: identity crisis. *Nature*. 2008 Mar ;451(7180):766–7. DOI: 10.1038/451766a

Wolinsky H. What's in a name? *EMBO reports*. 2008 Dec ;9(12):1171-4. DOI: 10.1038/embor.2008.217

6

Agricultural Knowledge Management: Research Trends in the Field of Agriculture and Allied Subjects

Dr. Madhav Pandey[1], Akanksha Pandey[2] and Dr. M.L. Lakhera[3] [1]

University Librarian Indira Gandhi Krishi Vishwavidyalya, Raipur, Chhattisgarh State

[2]Studies in Library & Information Science, Pt. Ravi Shankar Shukla University, Raipur, Chhattisgarh State

[3]Professor, Dept. of Agricultural Statistics & Maths
Indira Gandhi Krishi Vishwavidyalya, Raipur, Chhattisgarh State

ABSTRACT

The trends of prioritization in agricultural research on various disciplines have been studied and reviewed with the help of CAB Direct and various CD ROM Database CABI, AGRIS, AGRICOLA etc. Most of the papers were published by the researchers of faculty of agriculture, horticulture, agricultural engineering, veterinary and animal husbandry, fisheries. Prioritization based study of agricultural research in national and global level is also conducted and that indicate more balance research is needed. Recent time it is noticed that the research activities were increased in multifold due to availability of e-resources, various CD ROM database, CAB Direct, Science

Direct and Agricultural Consortium like Consortium for e-Resources in Agriculture (CeRA), J-gateplus etc. Support of Indian Council of Agricultural Research plays an important role to increased agricultural research activities through strengthening and development of library services in country.

Keywords: *Agricultural Knowledge Management, Agriculture, CAB Direct*

1. Introduction

India is the agricultural dominated country in globe. Agriculture is the main source of its economy. Our economy is depending on basis of crop productivity, it is not only provide food need but also provide employment opportunity to a very large scale. Therefore good quality research is the priority of our researcher. In this regards Indian Council of Agricultural Research doing tremendous job for agricultural development in India. The Indian Council of Agricultural Research (ICAR) is an autonomous organization under the Department of Agricultural Research and Education (DARE), Ministry of Agriculture and Farmers Welfare, Government of India. Formerly known as Imperial Council of Agricultural Research, established in 1929. The ICAR has its headquarters at New Delhi. The Council is the apex body for co-coordinating, guiding and managing research and education in agriculture including horticulture, fisheries and animal sciences in the entire country. With 101 ICAR institutes and 71 agricultural universities spread across the country this is one of the largest national agricultural systems in the world.

2. Agriculture Research in India

More than sixty All India Coordinated Research Project on Rice, Wheat, Chickpea, Maize, MULLARP, Barley Nematodes, Maize, Pigeon Pea, Arid Legumes, Perl millets, Small millets, sugarcane, cotton, groundnut, soybean, Rapeseed, mustard, oilseed, linseed, IPM, honeybee, forage, fruits, mushroom, vegetable, potato,tuber, palm, cashew, spice, medicinal plants, floriculture, soil test, fertilizer, water management, ground water, dry land agriculture, Agro meteorology, farming system, agro forestry, farm machinery, plat culture, animal energy, Plastic culture, goat pig, poultry, cattle, fisheries and thirty other network projects are running smoothly for agricultural research development under the umbrella of ICAR in India.

3. Materials and Methods

Research papers based review study will indicate the balance research in the field agriculture. Such type of research study is an important way to understand its past and according decide the future to meet the research on sustainable basis. The present review has been carried out to study the pattern of various aspect of agriculture with the help of CAB Direct, CD ROM Database CABI, AGRIS, AGRICOLA and ScienceDirect. Consortium for e-Resources (CeRA), Jgateplus databases were also used. We collect all the data and analyzed on excel and estimate the percentage and result which show in the paper.

3.1 CAB Direct

CAB Direct is CABI's online database platform, providing a single point of access to all of your CABI database subscriptions. CABI has worked with development partners from across academia and industry to develop the next generation of the CAB Direct platform. Designed around the way researchers work, CAB Direct has completely new features to help you get more out of the literature, a more intuitive user experience, and has a new look and feel. CAB Direct is the only online platform built specifically to help you get the most out of CABI's world class databases, CAB Abstracts and Global Health. Through CAB Direct we can access to: over 11.5 million bibliographic records, over 350,000 full text articles hosted by CABI and many other authoritative reviews, news articles and reports.

3.2 Consortium for e-Resources in Agriculture (CeRA)

The ICAR has provided Consortium for e-Resources in Agriculture CeRA under NAIP project from 2008 onwards. It is providing access to nearly 3700 journals in Agriculture and allied disciplines. The user ID and Passwords have been circulated to all the Colleges of the University, Research Stations and also to all the patrons of the University to utilize the e-Resources effectively.

4. Data Analysis

Table 1: Analysis of Research Papers Published in Agriculture and allied field Global and National Scenario

S.N.	Name of Discipline	Global Level		National Level	
		No. of Research Paper	Percentage	No. of Research Paper	Percentage
1	Agriculture	2102025	51%	87806	45%
2	Horticulture	254723	06%	28546	14%
3	Agril. Engineering	298641	07%	10292	05%
4	Veterinary and Animal Husbandry, Poultry, Fisheries	1206165	29%	47788	24%
5	Dairy	268033	07%	22767	12%
	Total	4129587	100%	197199	100%

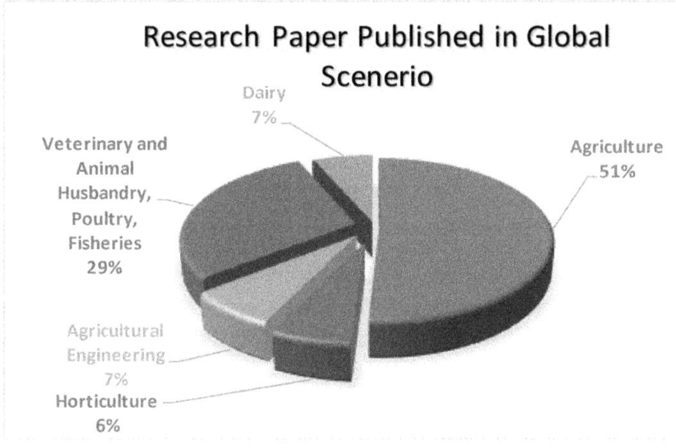

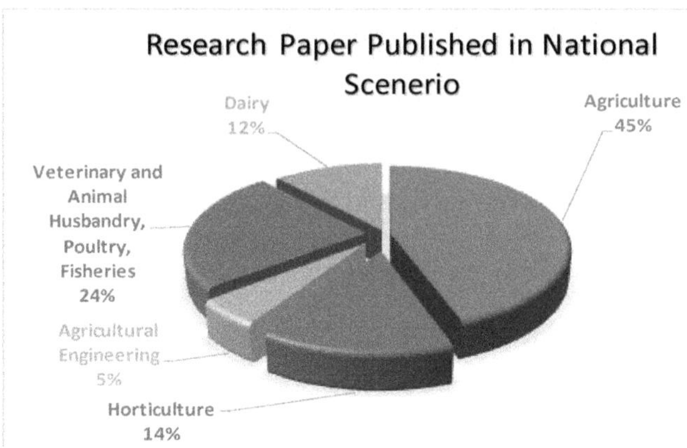

Table1 clearly indicates that the maximum research paper is published in agriculture and animal husbandry discipline on global level. The figure in case of India is also likely similar. However rest of the component agriculture engineering, dairy and horticulture also need to be strengthening in future to make more balanced and adoptable research for wider range of farmer community.

Table 2: Research Paper Published on Different Aspect of Agriculture

S.N.	Name of Discipline	No. of Research Paper	Percentage
1	Agronomy	102866	05%
2	Soil Science	399014	19%
3	Agricultural Entomology	274991	13%
4	Plant Pathology	261837	12%
5	Plant Genetics	182371	09%
6	Plant Breeding	253845	12%
7	Plant Biochemistry	100964	05%
8	Plant Biotechnology	174776	08%
9	Plant Physiology	172438	08%
10	Agril. Meteorology	10358	0.5%
11	Agricultural Extension	43722	02%
12	Agricultural Statistics	17338	01%
13	Agril. Economics	107505	05%
	Total	2102025	100%

Data show that in field of agriculture, meteorology and agriculture extension needs more concentration for further research because dissemination of knowledge will help to increase the production and productivity. This also helps for adoption of new technology in agricultural community.

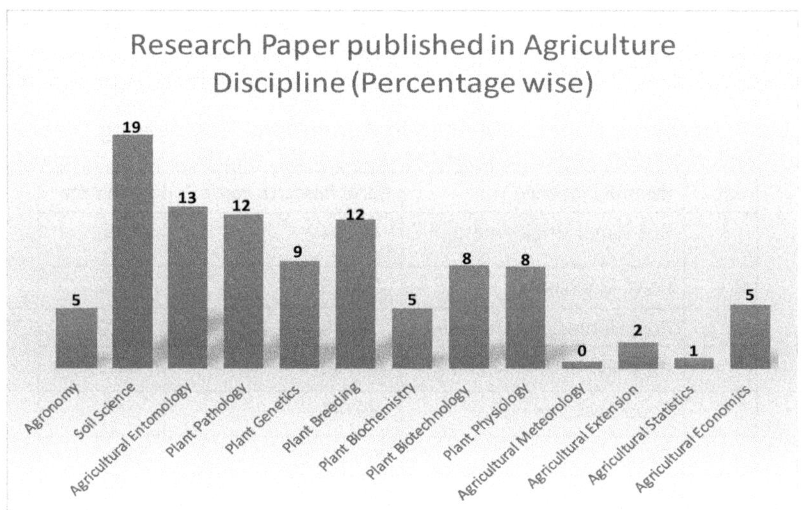

Table 3: Research Paper published in Horticultural Discipline

S.N.	Name of Discipline	No. of Research Paper	Percentage
1	Fruit Science	150616	58%
2	Vegetable Science	87321	34%
3	Floriculture	9100	04%
4	Ornamental Horticulture	11686	04%
5	Total	254723	100%

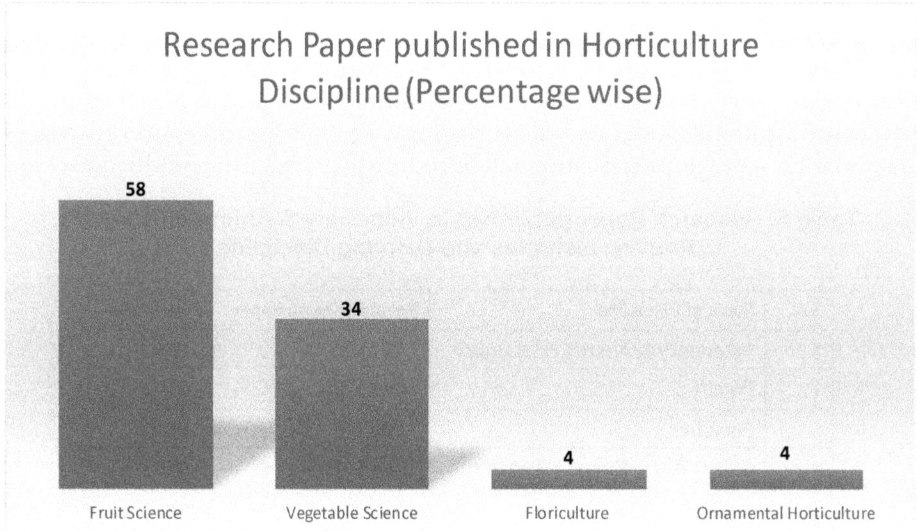

Data indicate that in field of horticulture maximum research paper published on fruit and vegetable sciences. Need is more concentration is given in remaining aspects i.e. ornamental horticulture and floriculture which can increase our export revenue.

Table 4: Research Paper published in Agricultural Engineering Discipline

S.N.	Name of Discipline	No. of Research Paper	Percentage
1	Soil Water Engineering	41804	14%
2	Farm Machinery	20355	07%
3	Post Harvest Technology	5556	02%
4	Agricultural Processing	150781	50%
5	Food Engineering	80145	27%
	Total	298641	100%

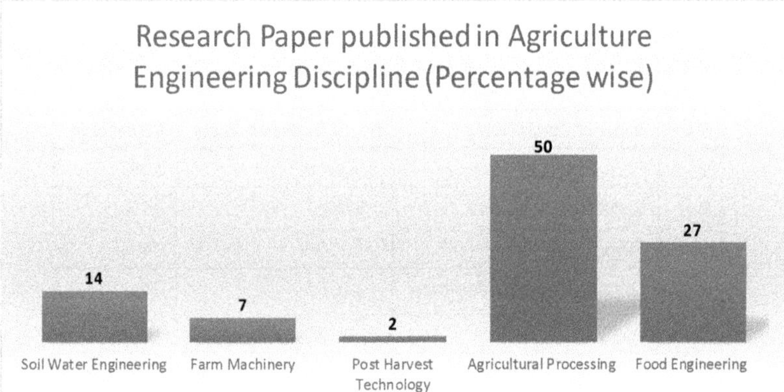

Table 4 and graph indicate that in agricultural engineering discipline maximum concentration is given on agricultural processing, food engineering and soil water engineering. But more intention is required on post harvest technology, if we concentrate on post harvest technology, so that the postharvest losses i.e. transportation, storage, preservation can be reduce to enhance the production of crops.

Table 5: Research Paper published in Veterinary & Animal Husbandry, Poultry, Fisheries and Dairying Discipline

S.N.	Name of Discipline	No. of Research Paper	Percentage
1	Veterinary & Animal Husbandry	851124	68%
2	Poultry	249315	20%
3	Fisheries	105726	08%
4	Dairy (Tech., Engg., Chemistry)	51814	04%
	Total	1257979	100%

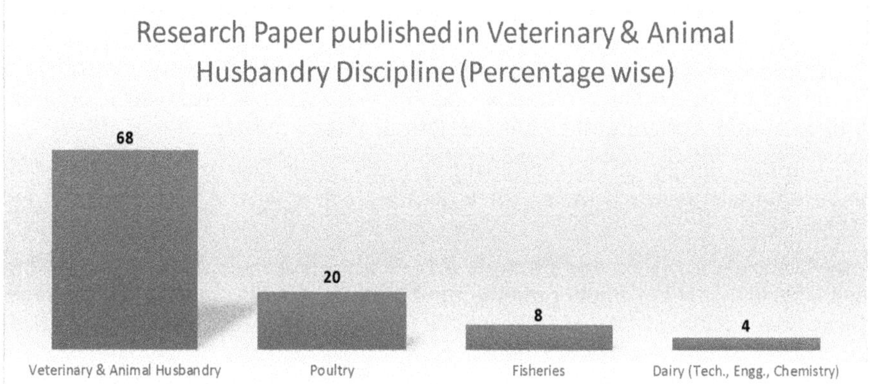

Table 5 and graphical representation shows that in veterinary & animal husbandry discipline maximum concentration is given on Vet. & AH.,but more concentration is needed in dairy research, which can enhance milk and milk product to improve nutrient value of society.

5. Conclusion

Most of the papers published by the agriculture scientists on various agricultural and its related aspects to enhance crop productivity for fulfills the food need of human society. It is understandable that need based priority is given by the researcher, because government and ICAR provide lot of support for researchers. Rest of the components of agriculture viz., horticulture, agricultural engineering, veterinary & animal husbandry, agricultural meteorology, ornamental horticulture, floriculture, post harvest technology, post harvest losses and dairy sciences are also need to be strengthened future to make more balanced research on various disciplines of agricultural sciences to ensure uniform development.

Reference

CAB Direct Online Database available at www.cabdirect.org

Consortium for e-Resources in Agriculture (CeRA) available at www.jgateplus.com

Indira Gandhi Krishi Vishwavidyalaya, Raipur, Chattisgargh available at http://www.igau.edu.in

Pandey, Madhav and Sangar, R.B.S. 2007. Documentation based review of Water Management studies in last three decades. Paper presented in South Asian Conference on Water in Agriculture: Management options for increasing crop productivity per drop of water. 15-17 Nov. 2007 organized by Indira Gandhi Krishi Vishwavidyalaya, Raipur. Abstracts: 299p.

Pandey, Madhav, 2013. Prioritization of Medicinal Plants Research in Global, National and Chhattisgarh Scenario. Paper presented and published in National Seminar on Non-Timber Forest Produce, Medicinal, and Aromatic Plants & Spices: Innovation for Livelihood Security, 23-24 Dec. 2013 organized by Indira Gandhi Krishi Vishwavidyalaya, Raipur. Proc.: 2-6

Pandey, Madhav, Jain, B.C., Watti, U.K., and Mishra, R.K.. 2004. Availability of literature on medicinal and aromatic plants in Nehru Library for identification and documentation. Paper presented and published in *National Seminar on Medicinal and Aromatic Plants-Biodiversity, Conservation, Cultivation and Processing* 26-27 February2005, organized by Indira Gandhi Agricultural University, Raipur. Abstracts: 64p.

Pandey, Madhav, Singh, A.P. and andSangar, R.B.S. 2009. Documentation based review of agronomical research in last three decades. *Lab to land*: An International Journal, 1(1):45-48

Pandey, Madhav,Watti, U.K. and Sharma, R.N. 2005. Analysis of biotechnology database available in the Nehru. Paper presented and published in *International Conference on Plant Genomic and Biotechnology 26-28 October 2005* organized by Indira Gandhi Agricultural University, Raipur. Abstract: 87-88.

7

Open Educational Resources and Open Access Journals with special Reference to Veterinary Science

Dr. K. N. Kandpal and S.S.Rawat

National Library of Veterinary Sciences,
ICAR-Indian Veterinary Research Institute,
Izatnagar –243 122, Uttar Pradesh
e-mail: kandpalivri@gmail.com

ABSTRACT

The paper focuses on Open Educational Resources and Open Access journals which are available in the field of veterinary sciences subject. There has been the significant revolution in the area of Information Communication Technology (ICT) and advanced communication with the help of networking. This paper also emphasizes the role of libraries and information professionals toward the present ICT era in exploring open educational resources (OERs) and open access journals (OAJs) resources available on the Internet for the use of students, scientists, and researchers in the area of veterinary sciences.

Keywords: *Open Educational Resources, Open Access Journals, Veterinary Science*

1. Introduction

The concept of open access came in to existence after the development of Information Communication Technology (ICT) during 1991. One of the most significant achievements in the area of information communication technology (ICT) and internet speed, it is the back boon of open educational resources (OERs) and open access journals (OAJs) with refers as digital content related to education and learning and research. The concept of OERs and OAJs are accessible to researchers and students without any cost and it is totally freely without any royalty, license fee etc. The OERs and OAJs can be used, reused, modified or adapted for new use without seeking any permission from copyright holder. All the users are got information without paying any charge and libraries and the library professional's role of provided the information to their clients quickly.

2. Definitions of Open Access

The term "open access" was first formulated in three public statements in the 2000s: the Budapest Open Access Initiative in February 2002, the Bethesda Statement on Open Access Publishing in June 2003, and the Berlin Declaration on Open Access to Knowledge in the Sciences and Humanities in October 2003. It refers to an unrestricted online access to scholarly research which is primarily intended for scholarly journal articles.

2.1 The Budapest statement defined open access as follows:

The Bethesda and Berlin statements add that for a work to be open access, users must be able to "copy, use, distribute, transmit and display the work publicly and to make and distribute derivative works, in any digital medium for any responsible purpose, subject to proper attribution of authorship." Despite these statements emerging in the 2000s, the idea and practice of providing free online access to journal articles began at least a decade before the term "open access" was formally coined. By 'open access' to this literature, we mean its free availability on the public internet, permitting any users to read, download, copy, distribute, print, search, or link to the full texts of these articles, crawl them for indexing, pass them as data to software, or use them for any other lawful purpose, without financial, legal, or technical barriers other than those inseparable from gaining access to the internet itself. The only constraint on reproduction and distribution, and the only role for copyright in this domain, should be to give authors control over the integrity of their work and the right to be properly acknowledged and cited. Computer scientists had been self-archiving in anonymous ftp archives since the 1970s and physicists had been self-archiving in arxiv since the 1990s. There are many degrees and kinds of wider and easier access to this literature.

2.2 What is Open Access Concept?

Conventional fee-based publishing models fragment worldwide scholarly journal literature into numerous digital enclaves protected by various security systems which limit access to only licensed users. What would global scholarship be like if its journal literature were freely available to all, regardless of whether the researcher worked at Harvard or a small liberal arts college, or he/she was in the United States or Zambia? What would it be like if, rather than being entangled in restrictive licenses that limited its use, journal literature was under a license that permitted any use as long as certain common-sense conditions were met? This is the promise of open access (OA). Needless to say, there are many challenges involved in trying to achieve this bold vision, and it is not embraced, or even viewed as being feasible, by all parties in the scholarly communication system. Without question, open access has significant implications for libraries, especially academic libraries. The summarized in single sentence is that Open access Resource is Digital, Online, Free of Cost, and Free of most of the Copyright and Licensing Restrictions.

2.3 Government of India Initiatives toward Open Access

The initiatives taken by the government of India in various ways i.e. sponsoring various programme through their various organization as well as private public partnership (PPP) with the help of Human Resource Development Ministry (HRD), University Grants Commission (UGC), National Knowledge Commission (NKC), Ministry of Agriculture and other professional bodies of various ministries like as INFLIBNET, ICAR, CSIR and other research institutes. UGC drafted a policy framework entitled "UGC (Submission of Metadata and Full-text of doctoral Theses in Electronic Format) Regulations, 2005" in respect of strengthen national capability of producing electronic thesis and dissertations and to maintain university level and framed the policy for creation of national level database of research output by the master and doctorate dissertations or theses works are initiative by the university or research institute.

2.4 Open Access Advocacy Initiatives at ICAR under NAARM

The National Academy of Agricultural Research Management (NAARM) organized a round table discussion on 'Prospects and Strategies for OER and Creative Commons in SAARC Countries' on 27th February 2013. NAARM and Commonwealth of Learning (COL), Canada, Jointly conducted a workshop on 'Emerging Practices of Open Educational Resources in Higher Education and Training' during 16-17 May 2013. DDG(Edn), ICAR address to the workshop and said '*Skill development through OERs in higher education using cutting edge technologies will give access and also improve imparting of higher education to larger population*' (according to The Hindu, 18th May 2013) and the major outcome recommendations of the workshops are as -

- Inculcate the culture of OERs through ICTs led intervention, capacity building and bringing attitudinal changes.
- Encouraging practice of Creative Common licensing by the scientists of Indian National Agricultural Research System (NARS)
- Developing the Community of Practice (CoP) among the stakeholders from various countries in SAARC region
- NAARM and Directorate on Knowledge Management in Agriculture in collaboration with ICRISAT should take lead in development of appropriate policy and also finding out the strategic interventions required to strengthen OERs and creative commons licensing.

A two-day workshop organized in association with Global Forum on Agricultural Research (GFAR), hosted at Food and Agricultural Organization (FAO), Rome on 'Open Access to Agricultural Knowledge for Inclusive Growth and Development' during 29-30 October 2014 and the major outcome as recommendations of the workshops are as -

- Well defined open access policies with necessary regulations for scientific and technological information and data taking into account privacy, security of information, IPRs and Career advancement policies.
- Action by DARE/ICAR, Department of Agricultural Extension and State nodal agencies for advocacy, awareness, capacity building and use of open data and information.
- Broadband 3/4G Connectivity for all India for use in e-agriculture along with e-governance, e-health, e-education etc. and affordable data connectivity to rural areas
- Proactive participation at national and international level regarding standards for Agriculture information and knowledge resources, developing agricultural knowledge grid, quality assurance, developing open access libraries at KVKs, institutional online repositories with central harvesting facilities, and partnerships with international platforms such as CIARD RING etc and creating local rings.
- Promoting and enabling appropriate use of mobile devices and Tablet PC by farmers, producers and those involved in Agri-food chains through training and learning such as through Farm Schools and village extension workers
- Curriculum changes such as integrating data sciences in library science and related disciplines, and informatics in agricultural and related sciences at graduate and post graduate levels
- 3. Open Access as Tool for Libraries to Dissemination of Information

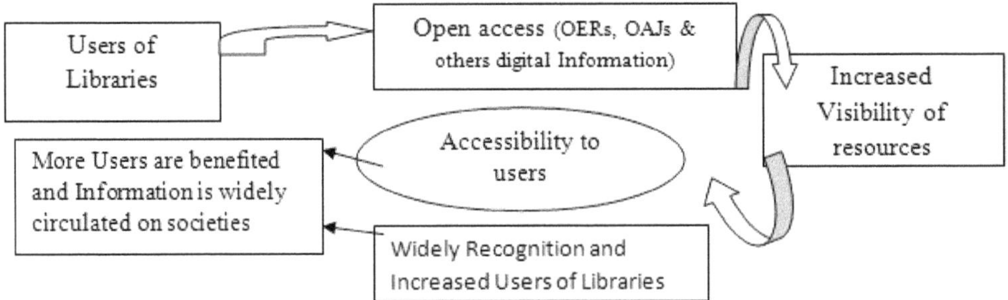

4. Role of Libraries and Library Professional in Open Access to their Clients

The role of libraries and its professional toward providing open access resources in their library is an important one. Open access does not require that libraries do anything for it to exist. The primary responsibility of the librarian and library professional are to provide cheep and best information to their clients in best possible time. From this perspective open access of information has the benefit and that too with no cost. For example, if a traditional journal becomes fully open access or a new open access journal fully substitutes for a conventional one, that is one less journal the library has to buy, and it can deploy those collection development funds elsewhere. After all the chances that libraries, mainly academic libraries, if not zero, will simply ignore open access materials is quite low. The lesson of other freely available Internet resources is that, disregarding of what libraries think, many users love them and make its good use to them to the prevention of conventional, vetted materials. So whether it was out of enthusiasm for new digital resources or out of a sense of obligation to steer users towards useful materials, libraries have increasingly considered that vast sea of Internet materials to be a source of materials that are a potential part of a redefined collection, one that primarily includes purchased and licensed materials, but also, through inclusion in digital finding tools and instruction, free Internet materials.

4.1 Libraries Can Be Digital Publishers of OA Works

A large number of libraries have now become digital publishers, primarily offering free/open access journals and institutional repositories. High quality free open source software is available to support digital publishing. Libraries are no longer simply consumers of scholarly information.

5. Difficulty of Assessing Open Access Impacts

As we saw in the earlier analysis of open access definitions, there is disagreement about whether the removal of price barriers is sufficient to achieve open access or whether, as is more commonly believed, the removal of permission barriers is also required. For example, many journals listed in the Directory of Open Access Journals (a widely recognized and used finding tool) do not remove permission barriers and neither do many e-print authors. In the self-archiving and open access journal discussions, we saw that, in reality, digital works commonly characterized as "open access" could be under a wide range of copyright and licensing arrangements. Looking solely at journals for a moment, the information environment is even more complex because there is a further distinction between free access to the entire contents of a journal and some subset of those contents.

1. Free Access journals (FA journals, color code: cyan): These journals provide free access to all articles and utilize a variety of copyright statements (e.g., the journal copyright statement may grant liberal educational copying provisions), but they do not use a Creative Commons Attribution License or similar license.

2. Open Access journals (OA journals, color code: green): These journals provide free access to all articles and utilize a form of licensing that puts minimal restrictions on the use of articles, such as the Creative Commons Attribution License.

3. Embargoed Access journals (EA journals, color code: yellow): These journals provide free access to all articles after a specified embargo period and typically utilize conventional copyright statements.

4. Partial Access journals (PA journals, color code: orange): These journals provide free access to selected articles and typically utilize conventional copyright statements.

6. Some most common online resources in the field of Veterinary Sciences

S.No	Name	Address
1	Open J-Gate/Online Journal Database	http://www.openj-gate.com/Search/QuickSearch.aspx or http://openj-gate.org/
2	Directory of Open Access Journals	http://doaj.org/
3	Acta Scientiae Veterinariae	http://www.ufrgs.br/actavet/
4	American Journal of Animal and Veterinary Sciences	http://thescipub.com/ajavs.toc
5	Animal Biology & Animal Husbandry - International Journal of the Bioflux Society	http://www.abah.bioflux.com.ro/
6	Institute of Genetics and Animal Breeding of the Polish Academy of Sciences	http://www.ighz.edu.pl
7	Annual Review of Biomedical Sciences	http://arbs.biblioteca.unesp.br/index.php/arbs
8	Bangladesh Journal of Veterinary Medicine	http://www.banglajol.info/index.php/bjvm
9	Bangladesh Veterinarian	http://www.banglajol.info/index.php/BVET/index
10	Japanese Association for Laboratory Animals (Experimental Animals)	http://www.jstage.jst.go.jp/browse/expanim
11	Global Veterinaria	http://idosi.org/gv/gv.htm
12	Human & Veterinary Medicine - International Journal of the Bioflux Society	http://www.hvm.bioflux.com.ro/
13	Iraqi Journal of Veterinary Sciences	http://vetmedmosul.org/ijvs/
14	Irish Veterinary Journal	http://www.irishvetjournal.org/
15	ISRN Veterinary Science Journal	http://www.isrn.com/journals/vs/
16	Italian Journal of Animal Science	http://www.aspajournal.it/index.php/ijas
17	The Journal of Poultry Science	http://www.jstage.jst.go.jp/browse/jpsa
18	Journal of Reproduction & Development	http://www.jstage.jst.go.jp/browse/jrd
19	Journal of Veterinary Medical Science	http://www.jstage.jst.go.jp/browse/jvms/-char/en
20	Journal of Veterinary Science	http://www.vetsci.org/
21	Livestock Research for Rural Development	http://www.lrrd.org/

22	Open Access Animal Physiology	http://www.dovepress.com/open-access-animal-physiology-journal
23	Open Journal of Animal Sciences	http://www.scirp.org/journal/ojas
24	Open Veterinary Journal	http://www.openveterinaryjournal.com/
25	Veterinary Research	http://www.veterinaryresearch.org/
26	Food Science and Technology International	http://fst.sagepub.com/
27	The Open Food Science Journal	http://www.researchgate.net/journal/1874-2564_The_Open_Food_Science_Journal
28	Livestock Research for Rural Development	http://www.lrrd.org/
29	Open Access Animal Physiology	http://www.dovepress.com/open-access-animal-physiology-journal
30	Open Journal of Animal Sciences	http://www.scirp.org/journal/ojas
31	Open Veterinary Journal	http://www.openveterinaryjournal.com/
32	Veterinary Research	http://www.veterinaryresearch.org/
33	Food Science and Technology International	http://fst.sagepub.com/
34	The Open Food Science Journal	http://www.researchgate.net/journal/1874-2564_The_Open_Food_Science_Journal
35	Foods — Open Access Food Science Journal	http://www.mdpi.com/journal/foods
36	American Journal of Food Science and Technology	http://www.sciepub.com/journal/AJFST#.Uhg0mdK38RM
37	The Journal of Microbiology, Biotechnology and Food Sciences	http://www.jmbfs.org/
38	Journal of Human Nutrition and Food Science	http://www.jscimedcentral.com/Nutrition/
39	Journal of Extension	www.jog.org
40	Elsevier Journals	www.journals.elsevier.com
51	Cambridge University Press	http://journals.cambridge.org/

7. Conclusion

A large number of scholars have openly supported open access as the high price of serial publications has hit the frequent access of same. They have either become open access advocates or have been swayed by its arguments. However the disciplines that are less dependent on journal literature have shown less enthusiasm and many scholars still have concerns about credibility issues associated with new digital publishing efforts and have not yet seen that the benefits outweigh the risks and costs in terms of time and effort (e.g., to create and deposit e-prints). Open access has struck a sympathetic cord in the library community, which has long suffered the debilitating effects of the serials crisis. However libraries have been somewhat cautious in their embrace of open access, uncertain about its destabilizing effects on the scholarly publishing system and its ultimate impact on their budget and operations. The open access movement is not the only potential solution to the serious problems that libraries face in the conventional scholarly communication system, but it is a very important one, and it does not require that other strategies be abandoned. Primarily as a result of the open access movement, there is now a rare opportunity to truly transform the scholarly communication system.

References

Charles W. Bailey, Jr. " The Spectrum of E-Journal Access Policies: Open to Restricted Access," DigitalKoans, 13 May 2005, http://www.escholarlypub.com/digitalkoans/2005/05/13/the-spectrum-of-ejournal-access-policies-open-to-restricted-access/. Note: Since the publication of this paper, College & Research Libraries has changed to an embargoed access journal.

Charles W. Bailey, Jr., " Early Adopters of IRs: A Brief Bibliography," DigitalKoans, 2 May 2005, http://www.escholarlypub.com/digitalkoans/2005/05/02/early-adopters-ofirs-a-brief-bibliography/.

Charles W. Bailey, Jr., "Evolution of an Electronic Book: The Scholarly Electronic Publishing Bibliography," The Journal of Electronic Publishing 7 (December 2001), http://www.press.umich.edu/jep/07-02/bailey.html.

Charles W. Bailey, Jr., "The Role of Reference Librarians in Institutional Repositories," Reference Services Review 33, no. 3 (2005): 266, http://www.digitalscholarship.com/cwb/reflibir.pdf.

Frederick G. Kilgour, "Historical Note: A Personalized Prehistory of OCLC." Journal of the American Society for Information Science 38, no. 5 (1987): 381

http://dspace.library.cornell.edu/handle/1813/62.

http://egj.lib.uidaho.edu/.

http://knowledgecommission.gov.in/downloads/documents/wg_open_course.pdf

http://scholar.lib.vt.edu/about/.

http://wiki.dspace.org/DspaceInstances.

http://www.doaj.org/.

http://www.insectscience.org/.

Jump up^ "Read the Budapest Open Access Initiative". Budapest Open Access Initiative. Retrieved 23 October 2013.

K Veeranjaneyulu (2015). Role of Agricultural Libraries in the Virtual Library Environment. NCALUC-2015, ICAR-Indian Veterinary Research Institute, Izatnagar (U.P.) ISBN: 978-810931095-0-2

Krista D. Schmidt, Pongracz Sennyey, and Timothy V. Carstens, "New Roles for a Changing Environment: Implications of Open Access for Libraries," College & Research Libraries 66, no. 5 (2005): 408-409.

Lon Savage, "The Journal of the International Academy of Hospitality Research," The Public-Access Computer Systems Review 2, no. 1 (1991): 54-66, http://info.lib.uh.edu/pr/v2/n1/savage.2n1.

Markin, Pablo (25 April 2017). "The Sustainability of Open Access Publishing Models Past a Tipping Point". OpenScience. Retrieved 26 April 2017.

Rathinasabapathy, G and L.Rajendran (2011). Mapping of Open Access Journals in Agricultural Science: A Study Based on Directory of Open Access Journals (DOAJ)

Rathinasabapathy, G, L.Rajendran and J.Arumugam (2012). Contribution of India for Open Access Journal Literature: A Profile Based on Directory of Open Access Journals (DOAJ). *In:* Conference Volume of International Conference on Trends in Knowledge and Information Dynamics (ICTK 2012), at Bangalore.

Raym Crow, A Guide to Institutional Repository Software, 3rd ed. (New York: Open Society Institute, 2004), http://www.soros.org/openaccess/software/; and SPARC, "Publishing Resources," http://www.arl.org/sparc/resources/pubres.html.

S K Soan (2015). OER and Open Access Movement in Agricultural Information Management: Current Status and Future Challenges for Indian NARS. NCALUC-2015, ICAR-Indian Veterinary Research Institute, Izatnagar (U.P.) ISBN: 978-810931095-0-2

Washington D.C. Principles For Free Access to Science: A Statement from Not-for-Profit Publishers," 16 March 2004, http://www.dcprinciples.org/statement.pdf.

www.ugc.ac.in/new_initiatives/etd_hb.pdf

8

Research Trends in Agriculture and Allied Sciences in India 2000–2007

Dr. SeemaParmar[1] and Dr. Rajive K. Pateria[2]

[1]Assistant Librarian, [2] Deputy Librarian
Nehru Library, Chaudhary Charan Singh Haryana
Agricultural University
Hisar- 125 001, Haryana

Abstract

The study has been carried out to analyse the research trends in the field of Agriculture and allied sciences during the years 2000-2007. Data was collected from a renowned database in the field of Agriculture sciences viz. 'Krishiprabha' to analyse and find out the trends, areas of research, research oriented universities etc. in the field of agriculture and allied sciences, so that the translucent scenario of agricultural research in the recent past may be known to agricultural community.

Keywords: *Research; Agricultural sciences; Doctoral dissertations; e-theses; e-repository; Krishiprabha; Bibliometrics.*

1. Introduction

The term "Research" is self explanatory as it clearly indicates that it is a continuous process and it walks over ages. Research may be conducted on every field of knowledge to explore different type of findings. Martyn Shuttleworth says that "the definition of research includes any gathering of data, information and facts for the advancement of knowledge" (website of Explorable). According to Cambridge Online Dictionary " research is a detailed study of a subject, especially in order to discover(new) information or reach a (new) understanding".

Doctoral research plays a very important role in the intellectual heritage of a country. It is the harbinger of scientific research in all domains of knowledge which initiates the researchers into a great scholarly journey to academia. The scholars pursuing higher education in agriculture and allied fields are awarded doctoral degrees on the basis of excellence in research as contained in their dissertations. Doctoral dissertations – the end product of doctoral research - are a rich source of original information on a given subject. The information enshrined in the dissertations needed to be fully exploited. Unfortunately, the dissertations received from the scholars catered to the needs of only local readers and research scholars. The research scientists in other institutes remained unaware of the similar work breeding elsewhere in the country. This ignorance led to duplication of work resulting into a huge waste of precious funds and human resources. Cases of plagiarism also came into notice at various places in the country. The dissertations remained confined only to the shelves of the libraries and the valuable information contained therein started losing its sheen with the passage of time. It is with this background that the idea of Krishiprabha was envisaged NAIP-ICAR through its ICT initiatives.

Krishiprabha is a big and only database of e-theses submitted to all Indian agricultural universities in the field of agricultural sciences. It was established mainly to develop, organize and sustain knowledge base of Indian Agricultural Doctoral Dissertations in digital form and make it accessible online. It was a systematic effort to develop and maintain a national repository of these resources, digitize them and make them accessible online that would facilitate their potential use for extracting maximum value from them.

2. Need of the study

The study has been carried out to make the agricultural academic community aware with the recent past trends in the field of agricultural research.

3. Objectives of the study

The study is done with following objectives:

- To study the recent past trends (2000-2007) in Agricultural research in India.
- To know about the areas of research in Agriculture and allied sciences.
- To identify the state wise, zone wise and university wise research output in Agriculture and allied sciences during 2000-2007.
- To get the discipline wise contribution of research in Agriculture and allied sciences.
- To know the guide wise contribution in research output in Agriculture and allied sciences.

- To know the year wise contribution in research output in Agriculture and allied sciences.
- To determine the diversity and trends of research in Agriculture and allied sciences

4. Methodology

The main source of data collection is "KrishiPrabha" database developed to create an Indian Agricultural doctoral Dissertations Repository. Data was explored and imported in to excel sheet to be filtered and analysed in different ways to find out the desired results.

5. Data Analysis

Collected data have been tabulated in different ways and then analysed using simple percentage method.

Table.1 State wise research outputs in Agriculture and allied sciences during 2000-2007.

Sr. No.	State	Research output	Ranking
1.	Andhra Pradesh	240	12th
2.	Assam	74	17th
3.	Bihar	101	15th
4.	Chhattisgarh	27	19th
5.	Gujrat	173	13th
6.	Haryana	810	1st
7.	Himachal Pradesh	335	10th
8.	J&K	100	16th
9.	Jharkhand	33	18th
10.	Karnataka	504	5th
11.	Kerala	161	14th
12.	Madhya Pradesh	16	21st
13.	Maharashtra	456	6th
14.	New Delhi	640	2nd
15.	Orissa	24	20th
16.	Punjab	446	7th
17.	Rajsthan	337	9th
18.	Tamilnadu	356	8th
19.	Uttar Pradesh	570	3rd
20.	Uttaranchal	508	4th
21.	West Bengal	243	11th
Total		6154	

Table shows the research contribution made by different Indian states in the field of Agricultural and allied sciences during 2000-2007. It is apparent from the table that Haryana state has contributed the highest research output (810) in agricultural sciences during 2000-2007 followed by New Delhi (640), Uttar Pradesh (570), Uttaranchal (508) and Karnataka (504), while least contribution was made by MP (16), Orissa (24), Chhattisgarh (27), Jharkhand (33) and Assam (74).

It is also clear from the table that each five universities have contributed less than 100 research outputs, five between 100-300, six universities contributed between 301-500 and five contributed more than 500 research output in the field of agricultural sciences.

Table.2 Zone wise research output in Agriculture and allied sciences

Zone	State	University	Research output
North	7	16	3409
East	5	6	475
West	3	10	966
South	4	8	1261
Central	2	2	43
Total	**21**	**42**	**6154**

The table shows zone wise states, universities and their research out in the field of agricultural and allied sciences. North zonal universities offered highest research output (3409) which was more than double that of second positioned south zonal universities (1261). Above table represents that North zonal universities' contribution is even more than sum total contribution of all other zonal universities.

It is also clear from the above table that north zone covers majority of states (7) and their agricultural universities (16). West zone covers only three states but 10 agricultural universities and made good contribution (966) in total agricultural research output.

Table.3 University-wise research outputs in Agriculture and allied sciences

Sr. No.	University	Research output
1.	*Anand Agricultural University*	108
2.	*Assam Agricultural University*	74
3.	*Bidhan Chandra Krishi Vishwavidyalaya*	188
4.	Birsa Agricultural University	33
5.	*Central Institute of Fisheries Education*	70
6.	*Chaudhary Charan Singh Haryana Agricultural University*	629
7.	*CSA University of Agriculture and Technology*	201
8.	*CSK Himachal Pradesh Krishi Vishvavidyalaya*	118
9.	*Dr Balasaheb Sawant Konkan Krishi Vidyapeeth*	44
10.	*Dr Panjabrao Deshmukh Krishi Vidyapeeth*	86

Sr. No.	University	Research output
11.	*Dr Yashwant Singh Parmar University of Horticulture and Forestry*	217
12.	*Govind Ballabh Pant University of Agriculture and Technology*	508
13.	*Guru Angad Dev Veterinary and Animal Sciences University*	8
14.	*Indian Agricultural Research Institute*	640
15.	*Indian Veterinary Research Institute*	235
16.	*Indira Gandhi Krishi Vishwavidyalaya*	27
17.	*Jawaharlal Nehru Krishi Viswavidyalaya*	16
18.	*Karnataka Veterinary, Animal and Fisheries Sciences University, Bidar*	7
19.	*Kerala Agricultural University*	161
20.	*Maharana Pratap University of Agriculture and Technology*	283
21.	*Maharashtra Animal and Fishery Sciences University*	21
22.	*Mahatma PhuleKrishi Vidyapeeth*	154
23.	*Marathwada Agricultural University*	81
24.	Narendra Deva University of Agriculture and Technology	127
25.	*National Dairy Research Institute*	181
26.	*Navsari Agricultural University*	36
27.	*Orissa University of Agriculture and Technology*	24
28.	*Pandit Deen Dayal Upadhaya Pashu ChikitsaVigyan Vishwavidhyalaya Evam Go Anusandhan Sansthan*	5
29.	*ANGRAU*	240
30.	*Punjab Agricultural University*	438
31.	*Rajasthan Agricultural University*	54
32.	*Rajendra Agricultural University*	101
33.	*Sardar Ballabhbhai Patel University of Agriculture and Technology*	2
34.	*Sardarkrushinagar-Dantiwada Agricultural University*	29
35.	*Sher-E-Kashmir University of Agricultural Sciences and Technology of Jammu*	25
36.	*Sher-E-Kashmir University of Agricultural Sciences and Technology of Kashmir*	75
37.	*Tamil Nadu Agricultural University*	237
38.	*Tamil Nadu Veterinary and Animal Sciences University*	119
39.	*University of Agricultural Sciences , Dharwad*	241
40.	*University of Agricultural Sciences, Bangalore*	256
41.	*Uttar Banga Krishi Vishwavidyalaya*	7
42.	*West Bengal University of Animal and Fishery Sciences*	48
Total		6154

University wise research output has been represented by the above table. Out of all agricultural universities, IARI and CCSHAU were the leading agricultural universities in providing highest research output (640 and 629 respectively) in the field of agricultural and allied sciences distantly followed by Punjab Agricultural University (438).

Out of total 42 agricultural universities, eight universities viz. Maharana Pratap University of Agriculture and Technology; UAS, Bangalore; UAS, Dharwad; ANGRAU, Hyderabad; Tamil Nadu Agricultural University; IVRI; YSPAU; CSAUAT have contributed research outputs between 200-300. Minimum research output less than 10 was made by five universities i.e Guru Angad Dev Veterinary and Animal Sciences University (8) Karnataka Veterinary, Animal and Fisheries Sciences University, Bidar (7); Uttar Banga Krishi Vishwavidyalaya (7); Pandit Deen Dayal UpadhayaPashu ChikitsaVigyanVishwavidhyalaya Evam Go Anusandhan Sansthan (5); and Sardar Ballabhbhai Patel University of Agriculture and Technology(2).

Table.4 Discipline wise research made in Agriculture and allied sciences

Sr. No.	Discipline	Research output
1.	*Agricultural Engineering and Technology*	157
2.	*Agriculture*	3919
2.	Basic Sciences	924
3.	Dairy Science	127
4.	Home Science	233
5.	Veterinary and Animal Sciences	794
Total		6154

Discipline wise research output in Agricultural and allied sciences has been presented in above table. It is obvious that most of the research work has been contributed in the discipline of Agriculture (3919). In other allied disciplines, more contribution was made by Basic Science (924) followed by Veterinary and Animal Sciences (794). Less research output was made by Dairy Sciences (127) followed by Agricultural Engineering and Technology (157). It is also clear from the above table that Agricultural research output is little less than double of all other disciplines' research output.

Table.5 Year wise research output in Agriculture and allied sciences

Year	Research output
2000	758
2001	765
2002	964
2003	771
2004	722
2005	829
2006	796
2007	549
Total	**6154**

The table shows the year wise research output in Agriculture and allied sciences. During the seven years period (2000-2007) taken for the study, maximum research output was made in the year 2002 as in that year a little less than thousand research theses (964) was awarded followed by 2005 (829) and 2006 (796).

Table.6 High Ranking of Universities and Guides in providing maximum research output

Sr. No.	Guide	University	Research Output
1.	*Choudhary, Maya*	Maharana Pratap University of Agriculture and Technology, Udaipur, Rajasthan.	13
2.	*Johri, B.N.*	Govind Ballabh Pant University of Agriculture and Technology, Pant Nagar, Uttranchal	12
3.	*Sawant, D.M.*	Mahatma PhuleKrishi Vidyapeeth, Rahuri, Maharashtra	11
4.	Kalia, Manoranjan	CSK Himachal Pradesh Krishi Vishvavidyalaya, Palampur, HP	11
5.	*Misra, R. L.*	Indian Agricultural Research Institute, New Delhi	10
6.	*Jaswal, Sushma*	Punjab Agricultural University, Ludhiana, Punjab	10
7.	*Jain, R.K.*	Chaudhary Charan Singh Haryana Agricultural University, Hisar, Haryana	9
8.	*Pant, R.C.*	Govind Ballabh Pant University of Agriculture and Technology	9
9.	*HariHar Ram*	Govind Ballabh Pant University of Agriculture and Technology	9
10.	*Gupta, H.C.L.*	Maharana Pratap University of Agriculture and Technology	9

The table shows the top ranked guides/advisor, their respective universities and number of research outputs they contributed the most during 2000-2007. Maya Choudhary of Maharana Pratap University of Agriculture and Technology has guided more research scholars in providing their research outputs followed by B.N. Johri of Govind Ballabh Pant University of Agriculture and Technology, Pantnagar, Uttranchal.

It is also interesting fact comes out from the table that out of top ten guides, three guide's fall from Mahatma Phule Krishi Vidyapeeth, Rahuri, Maharashtra and provided maximum research output (33 research theses). Similarly, next top three guides were from Govind Ballabh Pant University of Agriculture and Technology and provided 30 research theses as research output.

6. Major Findings

- Haryana state has contributed the highest research output in agricultural and allied sciences during 2000-2007 followed by New Delhi.
- Madhya Pradesh and Orissa state contributed least research output.

- North zonal universities offered highest research output which was more than double that of second positioned south zonal universities.
- North zonal universities' contribution was more than sum total contribution of all other zonal universities.
- North zone covered majority of states and their agricultural universities.
- Out of all agricultural universities, IARI and CCSHAU were the leading agricultural universities in providing highest research distantly followed by Punjab Agricultural University.
- Most of the research work has been contributed in the discipline of Agriculture while in other allied disciplines, more contribution was made by Basic Science followed by Veterinary and Animal Sciences while less research output was made by Dairy Sciences followed by Agricultural Engineering and Technology
- In Agriculture discipline, the research output is little less than double of all other disciplines' research output.
- During the seven years 2000-2007, maximum research output was made in the year 2002 as in that year a little less than thousand research theses was awarded followed by 2005.
- Maya Choudhary of Maharana Pratap University of Agriculture and Technology has guided more research scholars in providing their research outputs followed by B.N. Johri of Govind Ballabh Pant University of Agriculture and Technology, Pantnagar, Uttranchal
- Out of top ten guides, three guide's fall from Mahatma Phule Krishi Vidyapeeth, Rahuri, Maharashtra and provided maximum research output. Similarly, next top three guides were from Govind Ballabh Pant University of Agriculture and Technology.

7. Conclusion

There is an urgent need to know the agricultural scholars and faculty to know the recent and recent past trends in the field of their research so that they could get aware about their area of research in a proper way. Krishiprabha has really proved a boon for agricultural research community in the field of agriculture sciences. Theses of a particular period were archived in this Krishiprabha database. Afterward, this database on the name "Krishiprabha Community" is being updated in a separate agricultural e-repository "Krishikosh" by the respective universities. A few universities have started updating the community while others have yet not started. Updation of such kind of community is crucially required so that the same kind of studies could be done and researcher could become updated from their area of research.

References

Cambridge Dictionary Online http://dictionary.cambridge.org/ dictionary/english/research (Retrieved on 14.1.2016)

Creswell, J. W. (2008). Educational Research: Planning, conducting, and evaluating quantitative and qualitative research. 3rd Ed. Pearson, Upper Saddle River.

Krishiprabha Database. www.hau.ernet.in. Retrieved on 12-15 January 2016.

Mahapatra, R.K. and Sahoo, Jyotshna (2004). Doctoral dissertations in library and information science in India 1997-2003: A study. *Annals of library and information studies*. Vol. 51 (1): 58-63.

Mittal, Rekha (2011). Library and information science research trends in India. *Annals of library and information studies*. Vol. 58 (December): 319-325.

Parmar, Seema (2014). Krishikosh-A digital Repository in Agriculture and allied science. *Proceedings of ICLAM 2014:International conference on the convergence of Libraries, Archives and Museums* by National Institute of Fashion Technology,27-29 Nov,2014 at HauzKhas, New Delhi.

Shuttleworth, Martyn (2008). "Definition of Research". *Explorable*. Explorable.com. Retrieved on 14 January 2016 from http://www.explorable.com.

Wikipedia. http://en.wikipedia.org/wiki/Research (Retrieved on dated 14/1/2016)

9

Research Trends in Library and Information Science

Arundhati Kaushik

Deputy Librarian, University Library
G.B. Pant University of Agriculture and Technology
Pantnagar, Uttarakhand State

1. Introduction

Library and Information Science is one of the important professional subject has been growing tremendously in the last century and also in the present century. It is an emerging disciplines and being recognized worldwide as the e-learning resource centre which are one of the important components of any academic institutions like colleges, universities & research institutions including other institutions of higher learning. Library and e-Learning Resources Centre has been established as one of the vital component and hub of all the teaching –learning activities. Now the library and information science subject has been influenced by the wave of ICT and digital and web technologies which take it at new heights and leave behind the traditional librarianship. The younger generations are running after Information and Communication Technology application in library services and activities. In order to fulfill their needs recent trends in Library and Information Science Research (LIS Research) has been changed drastically towards ICT based research.

As learning, research and knowledge creation become more collaborative and dynamic, we need to make sure that the book is not our psychological trap. We need to consider how we tap into the content and repurpose it or bring it alive. How we can combine content with technology to provide a greater experience. Worldwide there is a growing demand for quality research performance measurement in Library and Information Science education. Library Professionals are keeping pace with the changing dynamics of the carious communities they serve. Library work creates value today in ways that are far more personal and collaborative. Library professionals need to be viewed as trusted advisors but trust grows only when we build relationships with our readers. The Library Professional who builds relationships – one who can coach, teach, or direct the readers to resources that support digital readiness – is the one who provides value today. They support the unique information needs of library customers by facilitating learning experiences. Today Library is less about what we have for people and more about what we do for (and with) people. This distinction is important because communicating the value of the Library Professional is the only way to ensure our future viability. Faster communication system and enhanced access to information bind countries, economies, and businesses in far more complex ways than we have ever conceived.

Author has compiled some of the researches on Library and Information Science subject after browsing Google and Google Scholar which would give at least some idea to the scholars in selecting their research topics. These researches are arranged in alphabetical order.

2. Research Trends

- A National Preservation Program for Agricultural Literature
- A Study and Practice on Agricultural Engineering Knowledge Database
- A Survey of the Awareness and use of Web 2.0 Technologies by Library and Information Professionals in Selected Libraries in South West Nigeria
- Access to Information: The Dilemma for Rural Community Development in Africa
- Accessing, Sharing and Communicating Agricultural Information for Development: Emerging Trends and Issues
- Agricultural Information and Knowledge Sharing: Promising Opportunities, for Agricultural Information Specialists
- Agricultural Information Dissemination using ICTs: A Review and Analysis of Information Dissemination Models in China
- Agricultural Knowledge Transfer in India: A Study of Prevailing Communication Channels
- Agricultural Library Information Retrieval Based on Improved Semantic Algorithm
- Application of Information Technology for Research in Tanzania: Feedback from Agricultural Researchers
- Australian Digital Library Initiatives
- Bridging the Digital Divide in India: Some Challenges and Opportunities

- Bridging the Information and Knowledge Gap Between Urban and Rural Communities through Rural Knowledge Centres: Case Studies from Kenya and Uganda
- CeRA – the e-Jourmal Consortium for National Agricultural Research System
- Challenges of Managing Information and Communication Technologies for Education: Experiences from Sokoine National Agricultural Library
- Cognition of Information Technology Consortium in University Library
- Content Management for Digital Delivery of Agricultural Information: Redefining Need of Libraries in the Context of Digitization of Theses and Research Reports
- Contribution of Television Channels in Disseminating Agricultural Information for the Agricultural Development of Bangladesh: A Case Study
- Development of Indian Agricultural Libraries from NATP to NAIP with Special Reference to Information and Communication Technology Applications
- Dial "A" for Agriculture: A Review of Information and Communication Technologies for Agricultural Extension in Developing Countries
- Digital Librarians and the Challenges of Open Access to Knowledge: The Michael Okpana University of Agriculture (MOUAU) Library Experience
- Digital Preservation and Permanent Access to Scientific Information : The State of the Practice
- Digitization of Theses and Dissertations: Status Quo India
- Disseminating and Promoting Agriculture Information through Library and Information Services in Ghana
- Documenting and Disseminating Agricultural Indigenous Knowledge for Sustainable Food Security: The Efforts of Agricultural Research Libraries in Nigeria
- Efficacy of Indexing and Abstracting Services in the Dissemination of Agricultural Information Resources in the Institute for Agricultural Research Library, Ahmadu Bello University, Zaria
- Evaluation of the Agricultural Information Service (AIS) in Lesotho
- Evaluation of Use of Consortium of e-Resources in Agriculture in Context of Kerala Agricultural University
- Factors Influencing Access to Agricultural Knowledge: The Case of Smallholder Rice Farmers in the Kilombero District of Tanzania
- Farmers Information Literacy and Awareness towards Agricultural Produce and Food Security: FADAMA III Programs in Osun State Nigeria
- Future of the Library and Information Science Profession: Special Libraries

- Globalization and Small-Scale Farming in Africa: What Role for Information Centres?
- Growth and Development of Agricultural Education, Research, and Libraries in India
- Growth and Development of Agricultural Education, Research, and Libraries in India
- HINARI: Bridging the Global Information Divide
- How can ICTs be Used and Appropriated to Address Agricultural Information Needs of Bangladeshi Farmers?
- ICT/ICM in Agricultural Research and Development: Status in Sub-Saharan Africa
- Implications of Library Consortia: How the Indian Libraries are benefitted?
- Improving Public Library Services for Rural Community Development
- Information and Communication Technology in Agricultural Development: A Comparative Analysis of Three Projects from India
- Information Communication Technology in Agricultural Development: A Comparative Analysis of Three Projects from India
- Information Literacy and International Capacity Development Initiatives in Life Sciences: AGORA, DARE, HINARI. ARDI (Research 4Life-R4L)
- Information Literacy Competency of Post Graduate Students at Haryana Agricultural University and Impact of Instruction Initiatives: A Pilot Survey
- Information Services Provision and User Satisfaction in Agricultural Research Libraries in Nigeria
- Information Technology Adoption in Libraries of Kerala: A Survey of Selected Libraries in Thiruvananthapuram
- Information Use of Software Packages in Nigerian University Libraries
- Initiatives that Centre on Scientific Dissemination
- Knowledge Creation and Flow in Agriculture: The Experience and Role of the Japanese Extension Advisors
- Knowledge Management Practices in Academic Libraries: A Case Study of the University of Natal, Pietermaritzburg Libraries
- Knowledge Translation in Agriculture: A Literature Review
- Major Agricultural Information Initiatives: With Emphasis on Developing Country Services
- Major ETD Initiatives in India
- Management of Indigenous Knowledge as a Catalyst towards Improved Information Accessibility to Local Communities: A Literature Review
- Management of Indigenous Knowledge for Developing Countries

- Managing Indigenous Knowledge for Sustainable Agricultural Development in Developing Countries: Knowledge Management Approaches in The Social Context
- Mapping of Studies on Employment Creation of Agriculture and Agro-processing in Kenya
- Mission, Initiatives, and Obstacles to Research in Agricultural Education: A National Delphi Using External Decision-Makers
- Networking of Agricultural Information Systems and Services in India
- Open Access Initiatives and Institutional Repositories: Sri Lankan Scenario
- Open Access Initiatives for Agricultural Information Transfer Systems in India
- Open Access Initiatives for Agricultural Information Transfer Systems in India
- Open Access Initiatives in India with Reference to Agriculture
- Realigning Library and Information Services and Extension Services to Sustain Agricultural Information Services in Zimbabwe: Case of Urban Farming
- Repository of Indian National Agricultural Research and Education System (NARES) – Open Access to Institutional Knowledge
- Researchers' Perspectives on Agricultural Libraries as Information Sources in Tanzania
- Restructuring LIS User Education Courses in Universities of Agricultural Sciences: A Study
- Revamping Agriculture Library and Information Services in India: Retrospect and Prospect
- Role of Agricultural Libraries in the Transformation of Agricultural Education and Sustainable Development in Nigeria
- Role of Agricultural Libraries in the Transformation of Agricultural Education and Sustainable Development in Nigeria
- Role of an Agricultural University Library in Technology Transfer in Agriculture
- Role of Information and Communication Technologies in Modern Agriculture
- Role of Information and Communication Technology (ICT) for Agricultural Development in Western Odisha
- Role of Information Technology in Agriculture
- Scholarly Publishing Initiatives at the International Rice Research Institute: Linking users to Publish Goods via Open Access
- Status of Rural Library and Information Services in Bangladesh: Directions for the Development

- Strategies of Indigenous Knowledge Management in Libraries
- Strengthening of Digital Library and Information Management in Agriculture (India)
- Strengthening of Digital Library and Information Management in Agriculture (India)
- Success through Working together: The U.S. National Agricultural Library's Approach to Cooperative Activities
- TEEL (The Essential Electronic Agricultural Library): A User Study
- The Impact of Information and Communication Technology on Research Output of Scientist in Two Selected Nigerian Agricultural Research Institutes
- The Impact of New Information Technology in the Developing Countries
- The Library's Role in Developing Information Literacy and Societal Growth
- The Library's Role in Developing Information Literacy and Societal Growth
- The National Agricultural Library's Database Online Documents Covering Water and Agriculture
- The New Development of Digital Libraries in China
- The Role of Information Services in Agricultural Development of Iran, A Progress Report
- The Role of Libraries in Supporting Agricultural Policy Research – Evidence from Selected University and Research Institute Libraries in Nigeria
- The Role of Research Libraries on the Information Needs of Agricultural Engineers (researchers) in Nigeria: A Case Study of National Centre for Agricultural Mechanization Library
- The Usage of e-resources among Agricultural Researchers and Extension Staff in Tanzania
- The Use of Libraries and Information Centres by Agricultural Researchers and Extension Workers in Zimbabwe
- The Use of Internet Services by Postgraduate Students for Research in Francis Idachaba Library, University of Agriculture, Makurdi
- Towards Effective Knowledge Management Practices for Agricultural Information Specialists in Tanzania
- Towards Indian Agricultural Information: A Need Based Information Flow Model
- Transformation of Indian Agricultural Libraries in a digital and Collaborative Era: A Case Study
- Use of Electronic Resources: A Study of Indian Agricultural Research Institute, New Delhi
- Use of e-Resources through Consortia – Delhi University Library System

- Use of ICT in Agricultural University Libraries in Western India: User Survey
- Use of Information Sources and Services in Library of Agriculture Science College, Shimoga: A Case Study
- Use of Library Information Technology Resources by Graduate Students of University of Agriculture, Abeokuta
- User Needs and Library Services in Agricultural Sciences
- Utilization of Automated Electronic Information Services: A Case Study at the University of Agriculture Library, Abeokuta, Nigeria

3. Conclusion

ICT and digital revolution paved the challenges for the researches in Library and Information Science (LIS) not only in India but also in the developed world. Increase in the use and access to Information and Communication Technologies (ICT) for LIS education and research is now more evident. Responsibilities of LIS departments and faculty are increasing to produce best LIS researches to lead the 21st century librarianship. The major responsibility of the LIS departments to groom LIS students in the philosophy, knowledge, and professional values of librarianship, as practiced in libraries and in other contexts, and as guided by the vision of the 21st century librarianship.

Reference (browsed and selected from)

www.google.com

Scholar.Google.com

10

Awareness for Accessing Digitalresources Subsribed by Thenmims (Deemed to be University) Mumbai, Maharashtra, India: A Study

Ravindra M. Mendhe[1] and Prof. Dr. Kishor N. Patil[2]

Deputy Librarian[1], NMIMS (Deemed to be University, Mumbai),Shirpur Campus
Maharashtra State. email: ravindramendhe@gmail.com
[2]Librarian, MPKV, College of Agriculture, Dhule, Maharashtra State

ABSTRACT

This paper highlights the important survey findings in respect of accessing digital resource awareness; digital resources use pattern and attitude of users towards the digital resources, which have been subscribed by the THENMIMS Deemed University. The Present study has been undertaken with a view to know the accessing pattern of digital resources awareness amongst the SVKM's NMIMS (Deemed to be University),Mukesh Patel Central Library, Shirpur campus Users for accessing to digital resources (IEEE, Science Direct, McGraw-Hill, Springer, ASME, ASCE,J-Gate,

ASTM Digital Library, Pro-Quest, JSTOR, NPTEL, DELNET, E-brary) subscribed by University. Digital resources and services are the prestigious issue of modern concepts. The present investigation is delimited to the randomly selected users of SVKM's NMIMS, Mukesh Patel Central Library, Shirpur Campus.Users who are regularly visiting Library for accessing to the digital resources subscribed by the University.

Keywords: *Digital Resources, User Awareness, NPTEL, DELNET, Information use pattern*

1. Introduction

The World Wide Web is constantly influencing the development of new modes of scholarly communication. The revolution of information and communication technology (ICT) across the globe haslot of changes in all departments. The advances in ICT have decisively changed the library and learning environment. Their efficiency for delivering goods is quite vast, as they successfully overcome the geographical limitations associated with the print media. With increase in the production of online information, users are in a better position of having up-to-date information to fulfill their needs. Such information is usuallysubscribed on IEEE, Science Direct, McGraw-Hill, Springer, ASME, ASCE,J-Gate, ASTME Digital Library, Pro-Quest, JSTOR, NPTEL, DELNET, andE-Brary,The Present study has been undertaken with a view to know the online Digital Resources awareness amongst the SVKM's NMIMS (Deemed to be University), Mukesh Patel Central Library, Shirpur campus Users for accessing to Digital Resources subscribed in University. Digital resources and services are the prestigious issue of modern concepts.

2. NMIMS (Deemed to be University), Shirpur Campus, Maharashtra

The NarseeMonjee Institute of Management Studies (NMIMS) (Deemed to be University) established in 1981 has over 13,000 students and 350 faculty members representing an eclectic mix of industry and academic experience in severed diverse domains of knowledge. Apart from its main campus at Mumbai, NMIMS has off-campus centers at Shirpur (Maharashtra), Bangalore and Hyderabad. In 2003, NMIMS established eight schools, offering programs across various disciplines including Management Technology, Science, Pharmacy, Architecture and Commerce offering graduate, postgraduate and doctoral programs.

Keeping in mind its concern for the society, NMIMS set up an off campus at Shirpur in 2007. This campus provides state-of-art infrastructure facilities to students while offering quality education. Spread over a land area of 50 acres, the campus comprises

- Mukesh Patel School of Technology Management and Engineering (MPSTME)
- School of Pharmacy and Technology Management (SPTM)
- MPSTME- Centre for Text Tile (CTF)

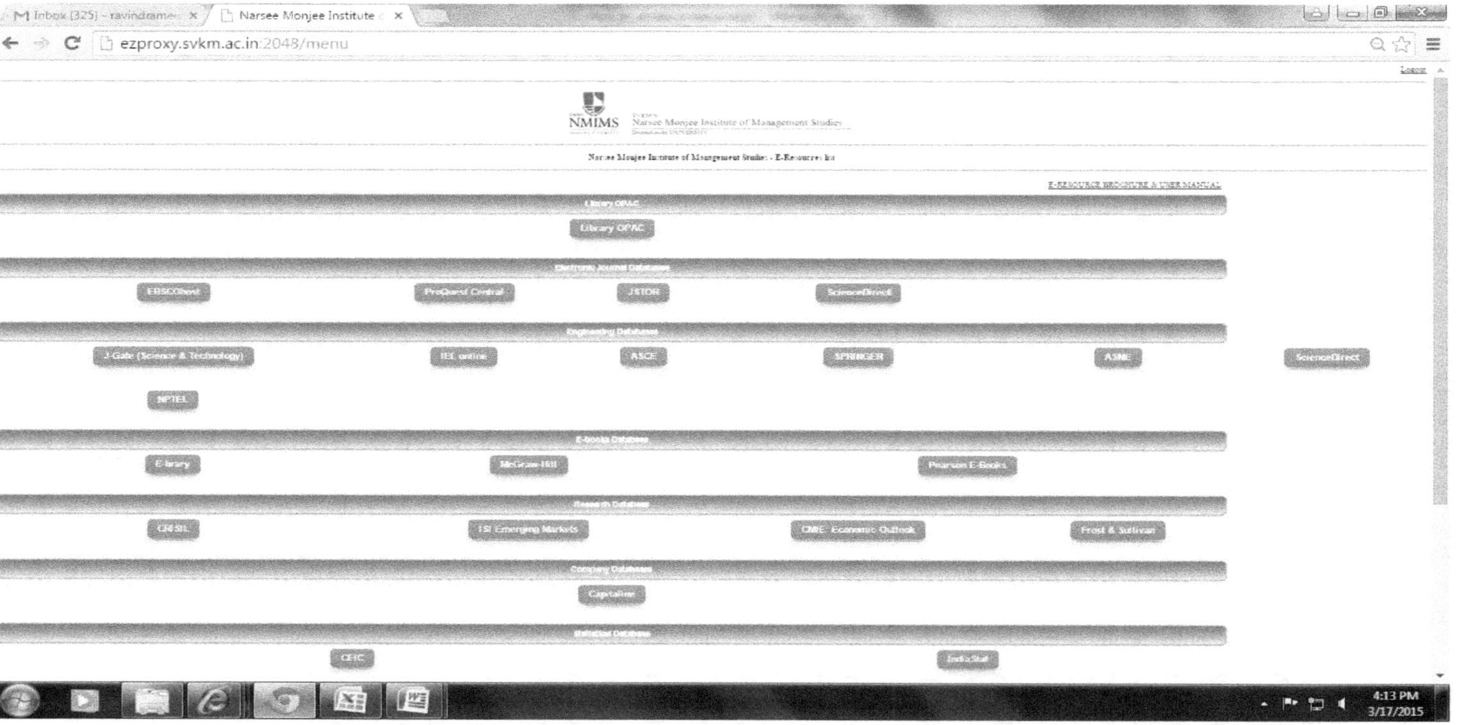

Figure 1: Digital Resources subscribed in NMIMS University

2.1. Mukesh Patel Central Library

Mukesh Patel Central Library at Shirpur campus is a valuable resource for all students and faculty members. It has a rich collection of 29500 books, 14various full text databases and 74,000 e-books from renowned publishers, NPTEL videos etc. Covering 45000 Sqft area, it provides access to valuable resources in Engineering, Management, Pharmacy, Computer Science, and Business Management.

3. Objectives

The purpose of this study is to analyzethe accessing to digital resources awareness amongst the SVKM's NMIMS (Deemed to be University), Mukesh Patel Central Library, Shirpur campus Users for accessing to digital Resources, the perceived impact of the digital resources on their academic efficiency and problem faced by them while using the online resources. The objectives can be summarized as:

- To know the different types of digital resources and services subscribed in the NMIMS Deemed to be University, Shirpur campus;
- To evaluate the usage pattern, frequency of usage, amount of time spent for accessing to Online Resources;
- To study the preference of the of Library users towards digital Resources services;
- To study the various suggestions put forwarded by the Central Library Users for the improvement of Internet Services being provided by NMIMS University.

4. LiteratureReview

A number of studies have been carried out on the Internet awareness of digital resources by teachers, students and research scholars of universities and research organizations. The present study is an attempt to establish and exhibit the Internet Awareness for Accessing to Digital Resources subscribed in SVKM's NMIMS (Deemed to be University) Shirpur Campus. Maharashtra, India.

The study by Sinha (2009) has also carried out a survey onspecialized group of samples who belongs toscientific disciplines (Participants of Workshop onBasic Science Research) in terms of ICT andInternet awareness and observed the similar trendsof finding towards awareness of ICT and Internetand utilization of E-Resources subscribed underUGC-INFONET Programme.Mendhe Ravindra, Dr.Pratibha Taksande &Dr.PreetiAgarwal (2008) conducted a study of the user of online resources by management faculty members affiliated to North Maharashtra University, Jalgaon to access the level of awareness and demand of web based learning environment among management information seekers. The major findings of the study revealed that majority of them used Internet for consulting technical reports regularly followed by e-journals, online database.

Naushad and Hassan (2003) have conducted a survey of the use of electronic information services by the users of IIT Library, New Delhi. The paper also examines the utilization and satisfaction level of users about Internet, CD-ROM database and other services provided by the library. Finally it highlights the suggestions made by the users for the further improvements of online services at IIT library Delhi.

Rahman and Ali (2010) conducted astudy on the access and utilization of the Internetbased library services subscribed to the faculty Members of Z.H. College of Engineering andTechnology, Aligarh Muslim University.However , it was also learnt that, there was no similar study conducted among faculty members, research scholar and students of MPSTME Library. Selvaraj and Rathinasabapathy (2015) have studied the information use pattern of students and faculty of Self-financing Engineering Colleges in Tamil Nadu and reported their findings including information usage pattern, preferred e-resources, etc. This study focus on awareness of digital resources subscribed in the NMIMS Deemed to be University.

5. Scope and Limitations

The scope of the present study confines to only the Mukesh Patel School of Technology Management and Engineering (MPSTME)Engineering faculty members, research scholar and Students of the NMIMS, Shirpur Campus, Maharashtra, India.

6. Methodology

In the Present study the questionnaire method has been chosen to collect information from the respondents. A well structured questionnaire covering all the facets of the topic has been prepared and distributed to faculty members, ResearchScholars, selected Under-Graduate / PostGraduate Students of NMIMS Deemed to be University, Shirpur Campus, Maharashtra.Altogether 210 numberof library users were selected for the present study.The questionnaire, was designed keeping in view evaluating the awareness for Accessing to Digital Resources,Usage Pattern of the Library Users.

7. Data Analysis and Interpretations

7.1. Questionnaire Distributed and Responded:

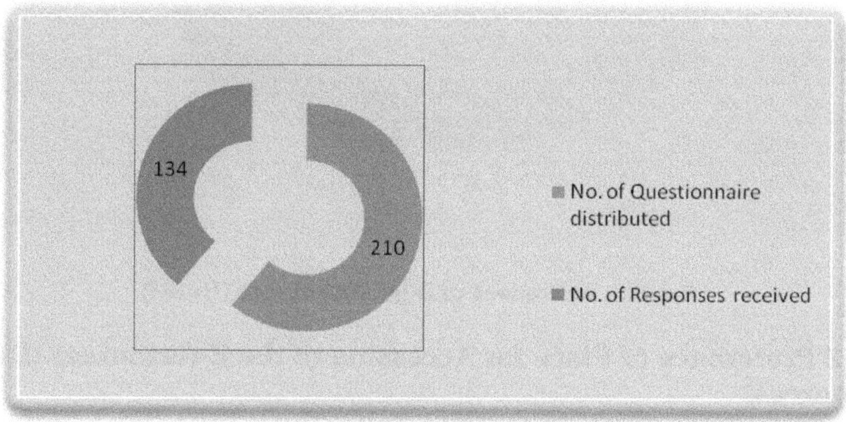

Graph 1: Response Rate of Users

Graph and Table No.1 Show the Questionnaire was distributed to 210 respondents and out of which 134 (63.88%) respondents have respondents.

Table 1: Response Rate of Users

Respondent	No. of Questionnaire distributed	No. of Responses received	%
Faculty Members	75	62	82.66
Research Scholars	50	31	62
Selected UG / PG Students	85	41	37.64
Total	210	134	

7.2 On-line Access to Digital Resources subscribed in NMIMS (Deemed to University)

7.2.1 Awareness of Digital Resources subscribedin NMIMS(Deemed to University)

Graph 2 show that out of 134 respondents, a maximum of 112(83.58 %) respondents are aware of the E-Resources subscribed in NMIMS (Deemed to University) for access to more than 14 Online Database subscribed whereas 22 (16.41%) are not aware of the online resources.

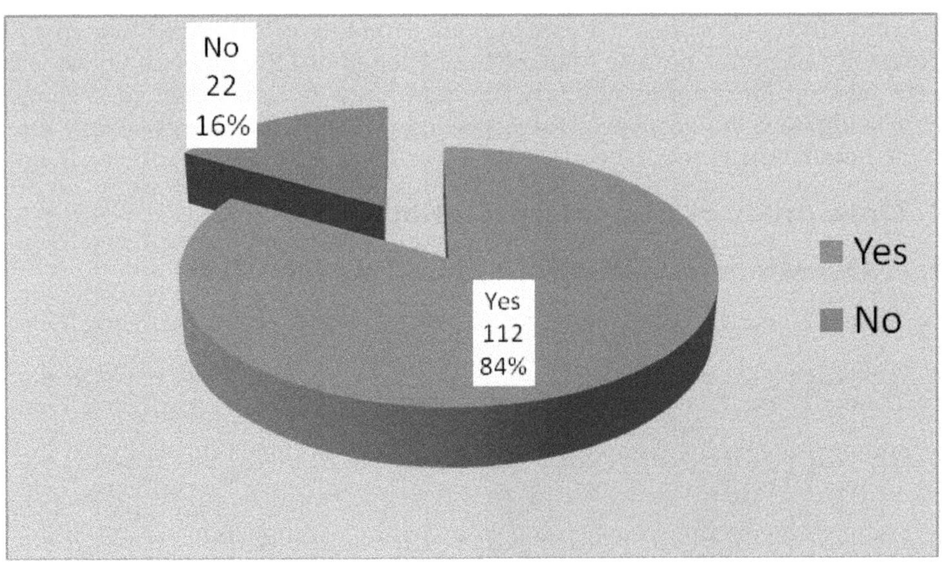

Graph2: Awareness of DigitalResources (N=134)

7.2.2 Preference of Place for Accessing to the E-Resources (Digital Resources)

Graph 3reveals that out of 134 respondent maximum numbers of respondents 90 (42.85%)prefer their own laptop to access E-Resources where as Computer Centre

34 (16.19%) is second preference and few respondents used to access E-Resources from office10 (4076%).

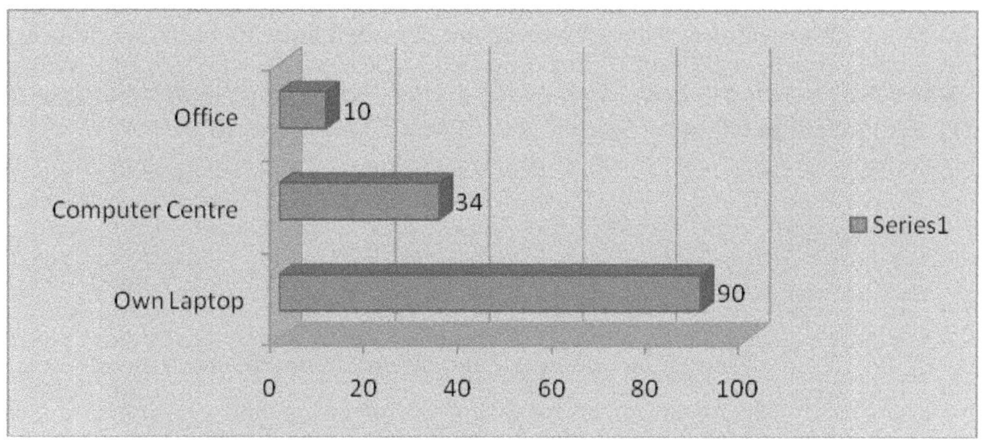

Graph3: Preference of Place of Accessing to E-Resources (N=134)

7.2.3 Frequency of E-Resources (Digital Resources) Access

Generally faculty and researchers scholar are more interested in accessing to e- journals for their research and teaching work, in this study author has tries to know the frequency of the respondent to access e-journals from varies E-Resources subscribed by University. Out of 134 respondents, a maximum of 68 (50.74%) respondents are accessing e-journals on daily basis and 31 (23.19%) respondents access e-journals on bi-weekly basis whereas 18 (13.43%) and 11 (8.20%) respondents preferred to access e-journals on weekly and fortnightly. While only few 6(4.47%) respondents are not regular in accessing the facility of e-journals/e-resources subscribed by University.(Show Graph no. 4).

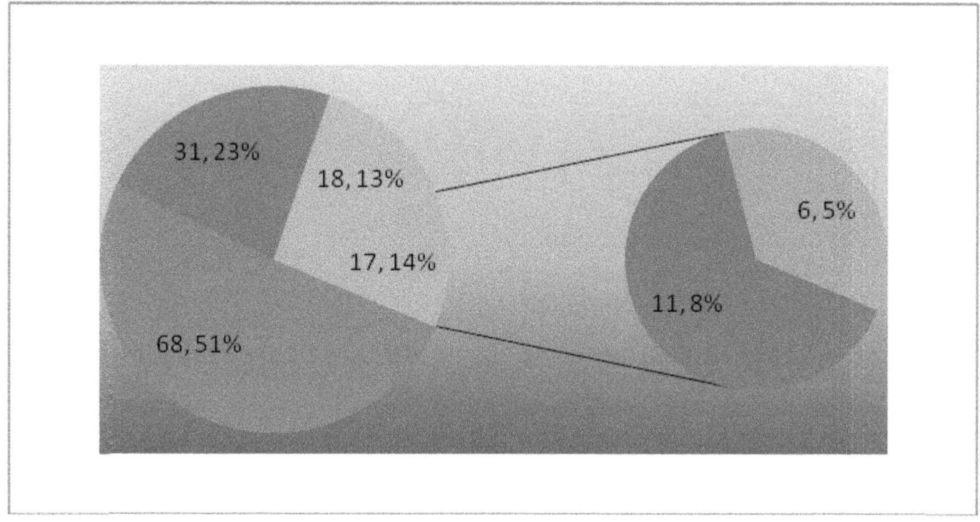

Graph 4:Frequency ofDigital Resources Access (N=134)

7.2.4 Rating of DigitalResources subscribed by NMIMS University

Graph 5 indicates that a maximum of 108 (80.59%) respondents have rated IEEE is highly useful which is followed by Science Direct 63 (47.01%) , E-brary 62(46.26%), Springer 47(35.07%),McGraw-Hill 34 (37.61%),ASME 31(23.13%),ASTM 26(19.40%),NPTEL 22 (16.41%),ProQuest 21(15.67%)ASCE 19(14.17%), J-Store 15 (11.19%),J-Gate 14 (10.44%), Scopus 12 (8.95%), DELNET 11(8.20%)

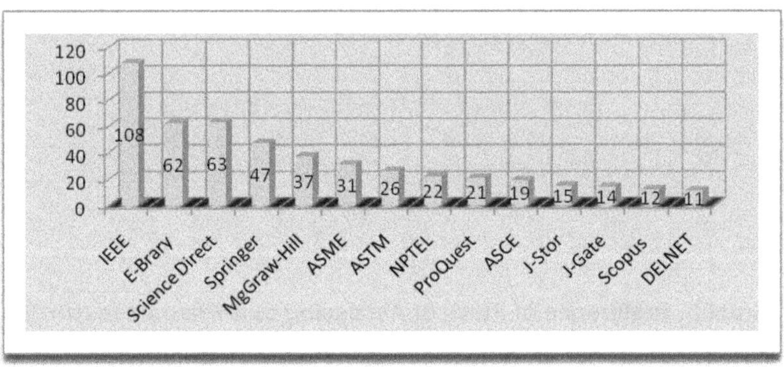

Graph 5: Rating of Digital Resources subscribed in NMIMS University (N=134 Multiple Choice)

7.3 Obstacles for On-Line access to Digital Resources

The surveys results show that in Graph -6 reveal that there is lot of obstacles while accessing to Digital resources subscribed in NMIMS University. Out of 134 respondents, a maximum of 63 (45.22%) respondents are complain about the slow speed of Internet and 29 (21.64%) respondents say that there is overload information on Internet, whereas 17 (12.68%) respondents find difficulty in the particular website of e-resources. About 14(1.44%) respondents say electricity/ power problems and 13(9.70%) say Wi-Fi connectivity problem.

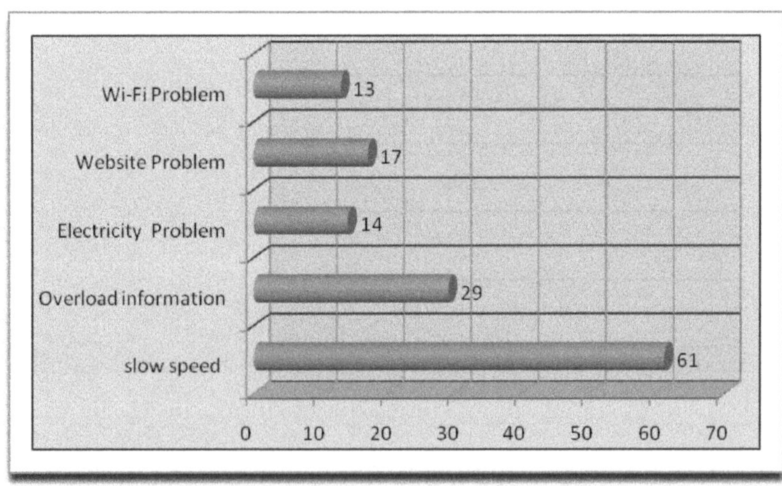

Graph 6: Obstacles for On-Line access to Digital Resources (N=134)

7.4 Suggestions from Respondents

The following suggestions given by the respondents.

- Subscribed IEL-IEEE online in a multiple user
- Installing High Capacity UPS
- Provision of separate terminals for faculties
- Need Dedicated Lease Line for E-Resources
- Printing facilities should be made subscribed to Library Users

8. Conclusion

The study sought to examine "Awareness for Accessing to Digital Resources subscribed in SVKM's NMIMS (Deemed to be University) Mumbai, Maharashtra, India: A Study", by taking samples from various faculties, Research Scholar and students of NMIMS. Most of the objectives are met within the results, awareness for accessing to digital resources subscribed SVKM's NMIMS is likely to differ from one faculty to another and among levels of status such as faculties, research Scholar andStudents.It is observed that the availability of digital resource on the campus is almost sufficient for all the existing disciplines but the infrastructure to use these resources is not adequate and can hinder the ability to meet the requirements of users.It may be concluded that it is on line resources, which make our life easy, if it is used in proper way for the benefit of the society.

References

Sinha, Manoj Kumar .(2009). A Study on ICT and Internet Awareness amongst the Research Scholars and University Teachers: A Case Study. INFOLIB ,**2**(1-4), 8-18.

Mendhe, Ravindra, Taksande, P G and AgrawalPreeti,(2008) Use of Digital Resources by Management Faculty Members affiliated to North Maharashtra University, Jalgaon ,In NACLIN – 2008, 11th National Convention on Knowledge, Library and Information Networking, 4-7, Nov.2008.

Naushad P.M. and Hassan. MdEshan (2003), The use of electronic services at IIT Library Delhi: a study of users' opinion. IASLIC Bulletin. 48(2), 71-82.

RahmanWahidur and Ali ,Amjad. (2010). Utilization of Internet –Based Library and Information Services by the Faculty Members of Engineering Colleges of Aligarh Muslim University, Aligarh: A Case Study. In: Reengineering of Library and Information Services at Digital Era: Proceedings of 7th Convention PLANNER-2010, 18-20 February 2010 (Eds: Jagdish Arora et. al.,)

Selvaraj, A. D., and G. Rathinasabapathy (2016). A Study on Electronic Information Use Pattern of Faculty Members of Self-Financing Engineering Colleges in Tiruvallur District, Tamil Nadu. Asian Journal of Library and Information Science 6.3-4 (2016): 31-40.

Selvaraj, A. D., and G. Rathinasabapathy. (2015). Information Use Pattern by Students of Self-Financing Engineering Colleges in Tiruvallur District, Tamil Nadu: A Study. Journal of Library, Information and Communication Technology 6.3-4 (2015): 45-54.

11

Role of Information Technology in the development of Mahatma Phule Krishi Vidyapeeth Library

Shri.Kapil Ramesh Kalhapure and Shri. Dinesh DattujiWakode

UniversityLibrary
Mahatma Phule KrushiVidyapeeth, Rahuri, Maharashtra State
e-mail: kapilkalhapure26@gmail.com

ABSTRACT

Application of information technology in libraries is increasing steadily and most of the libraries especially academic libraries are now using many IT tools and techniques to augment its services towards the user community. This paper attempts to explore the role information technology applications in Mahatma Phule Krushi Vidyapeeth University Library and its impact on the library user services.

Keywords: *Information Technology, Library Automation, E-Books, E-Theses, E-Services, Internet, MPKV*

1. Introduction

In this age of Information Technology, tremendous changes have been taken place in the field of Library Science. After invention of printing technology, the knowledge and information were being preserved in the form of print media. But now-a-days it can be stored electronically. The concept of library is not only to procure, preserve and make available information or knowledge to the seeker or reader in the form of books, journals etc but also electronic form. The revolutions of information and communication technologies have made the transfer of information from one place to another place easily and it can be accessed from any corner of the globe with help of computer networks which is called as Internet.

2. Information Technology (IT) - Definition

We use the term information technology or IT to refer to an entire industry. In actuality, information technology is the use of computers and software to manage information. In some companies, this is referred to as Management Information Services (or MIS) or simply as Information Services (or IS). The information technology department of a large company would be responsible for storing information, protecting information, processing the information, transmitting the information as necessary, and later retrieving information as necessary.

3. IT in Library and Information Science

ICT has changed the role of university libraries. One of the interesting effects is a much stronger orientation towards the university's strategic processes. Libraries are playing a supporting role in IT offers a range of opportunities to library and information professionals. These include providing new services and resources, enhancing the role of the information centre within the organization, and careerdevelopment. IT and teaching is unlikely to replace face-to-face training and education. It is rapidly becoming an important additional delivery method and it offers new learning opportunities. IT offers the opportunity for information workers across different countries to work-together and construct their own professional knowledge through virtual communities of practice. It requires information workers to develop new skills in the design and development of IT materials and programmes.

4. Need of Digital Libraries

a) *Storing Capacity:*

   A very large amount of information can be stored in e-document in comparison to printed document.

b) *Easy searching:*

   As e-documents fundamentally are text based files, so any one can search easily his needy material without reading the whole document.

c) *Facility of downloading*:

   Subject catalogues of e-documents are available for down loading in personal computers of the users.

d) *E-document reader :*

   The subject catalogues of e-documents are dedicated to hardware progress, in which the screen in of high quality that has special power of reading e-documents.

e) *Access on web :*

E-documentsare available on websites of service providers. So we can access these websites by paying some money.

f) ***Facility*** of copying :

As the computer is also linked with high quality printer so any one if wants can get the copy of any page of the documents through printer.

Let us discuss the utilization of newer technology in the development of MPKV library.

5. MPKV Library

5.1 MPKV Library Knowledge portal

MPKV Library knowledge portal is an important tool designed and developed for the benefit of the students and teachers in the University. The resources collected for the portal are classified as Online catalogues of holding, Online journals and databases subscribed. Useful websites on agriculture and allied subjects and e-resources like databases created by the library. Thc portal is hosted on local intranet of the University. General information can be accessed directly at URL: http://library/home.htm. However, the retrieval of databases or e-resources are restricted to the authentic members are restricted to the authentic members working in Central Campus of MPKV, Rahuri. Username and passwords are provided to the members.

5.2 Security in MPKV University Library

The library systems coordinator and the manager of library administration services provided the current status of various security controls for Mahatma PhuleKrishi Vidyapeeth, University library. MPKV library have security policies and procedures for library staff. Mahatma Phule Krishi Vidyapeeth, University Library tries to protect a great number of electronic resources. Library tries to protect books, various multimedia, equipment, and online digital resources. They also protect our human resources, of course. Their server is located in another building for security protection, All daily activities are backed up on this server, The tapes are then placed in their vault, which is protected with a fire-retardant unit called the Halon system. Books that are checked out must be desensitized before they can leave the building. Mahatma Phule Krishi Vidyapeeth, University library keep all their rare books in their vaults, which are climate-controlled. The entire basement of the building is on a pressurized sprinkler in case of fire. Mahatma Phule Krishi Vidyapeeth, University library have 16 security cameras located on the inside of the building. They have security cameras and monitors, motion detectors, coded locks, and alarms throughout the building. The Mahatma Phule Krishi Vidyapeeth,University library' firewall, anti-virus, and Microsoft Windows Active Directory LDAP authentication to their desktops and laptops. They use a proxy server authenticated to the campus Microsoft Windows Active Directory LDAP services for electronic resources.

5.3 Institutional Repository

Institutional Repositoryis an essential tool for categorization and dissemination of information and to enrich teaching, learning and research in the digital age. Therefore MPKV library has planned to create or to develop al repository for the Institutional Research scholars, faculty members, Graduate/post graduate students and other members of university library and provide them an opportunity to access, communicate and publish their research. We have identified the materials for an Institutional Repository such as newspapers (born digital), university publications, Departmental materials, Conference proceedings, etc., still these materials are not added in the repository buy the kinds of materials that are planned for inclusion in the near future. Library has developed a list of content currently present in the repository.

5.4 Digital Library of MPKV

The University Library, MPKV, Rahuri is using *KOHA* Open Source Software for digitization of documents with OPAC search module. At present MPKV library is in second phase of computerized library management systems but moving towards fully digital steadily. The hardware available in MPKV library is: Computer: 81, Printers: 6, Scanner: 3, Barcode Printer: 2, Barcode Reader: 2, Pen Drives: 12(16GB/32GB each) etc.

5.5 E-Services

In this age of electronic era, the face of library has totally changed with the advent of information Technology. Modern techniques to search information such as internet, CD ROM Databases (23), Online journals (5153), Online e-books(163) are introduced in the library. The revolution of information and communication technologies have made the transfer of information from one place to another place easy and can be accessed from any corner of the globe with the help of computer networks which is called as internet. The library has accepted this challenge to introduce these modern techniques to provide IT-based facilities such as. CD ROM Search. e-books, e-journals, on line Databases, E-mail etc. e-consortia.

5.6 Internet Facility

The University Library has developed a well qeuipped IT infrastructure with 70 computers with high end configuration for surfing the internet. Well furnished internet Call is provided to facilitate this service. The library has V-SAT connectivity with 94 mbps bandwidth speed.

5.7 CD Rom Databases/On-line Database

The University Library has 21 CD ROM Databases on agriculture and allied subject. The members of the library can retrieve references from these databases through intranet of the University.

5.8 E-journal/on-line E-journals

The university Library has some e-journals viz. Soil Science Society of America Journal, Vol.1-64 (1936-2000) and ActaHorticulturae (112Nos) on CDs and Library members can use is through intrenet of the University. The University Library has available CD ROM Database (1972-2012) and e-Journals 5 CDs in CD Form.

The University Library has recently subscribed Online Foreign Journals 5153 and 16 Indian journals for the year 2016 on agriculture and allied subjects. These can be accessed through internet at http://indianjournals.com. This facility is made available to the Library members only.

5.9 E-books

The university Library has 163 e-books (CRC Netbase.com) on agriculture and allied subjects Apart from this the statistical database viz. and 50 e-Books in CD Form. Foreign Trade Statistics oflndia (48 CDs 2003-2006) and Monthly Statistics of Foreign Trade of India (12 CDs 2005) are purchased and made available to the readers in library premises only.

5.10 ETAD (Electronic Thesis Abstracts Database)

Thesis or dissertation submitted by the students to the university content valuable findings. The MPKV Library has 9059 thesis and dissertations of M.Sc. (Agri), Ph.D, B.Tech(Agril. Engg), M.Tech (Agril.Engg) in its collection. To create database of the same a software was developed from local professional with eighteen fields. The scanning of abstracts was done by HP Scanner by using Precision Scan Pro 3.02 program. The Boolean operator such as 'and' or 'not' are used in search module. The database created is presently available at http://mpkv.mah.nic.in. One can now retrieve this database from any corner of the globe free of cost.

6. Conclusion

With advanced information technology, almost every library is changing from traditional library to digital library, So most of the people can grasp expected knowledge from any corner of the world and utilize it for his/ her development. MPKV library also has fully utilized new information technologies to provide better library services to its users. It is getting one of the best example of ideal agricultural university libraries. Implemented mode IT tools and techniques to render its quality services to the users.

References

Definition of digital library (online) availability athttp://adler.lib.muohio.edu/pipermail/disc-1/2002-november/000079.html

Kuzma, J. (2010) European Digital Libraries: Web security Vulnerabilities, Library Hi Tech 28 (3)

Mahatma Phule Krishi Vidyapeeth, University Library : At a Glance.

PremSingh (2004) Library Databases: Development and Management. Annals of Library & Information Studies, Vol. 51 (1)

Rajeshkar, T. B. Institutional Repository Software: Features and Functionality Bangalore : NCSI, IISC,2005

Rajeshkumar, G., D. L. N. Chary, K. Indrasena Reddy (2008), Need of Digital Libraries in modern era, wisdom publication, Delhi.

Somasundaram, M. (2005). Distance Learning to E Learning transformation from traditional to virtual education. IJTD 35 (1)

Sonkar, S. K. and Mahawar, K. L. Knowledge Portal : An innovative approach for the Libraries in knowledge Library and Information Networking NACLIN. 2007; 185-192.

12

A Bibliometric Analysis of contributions in the Journal of Dairying, Foods and Home Sciences (2003–2012)

Shanta S, Betageri[1] and Girish.S. Bhogashetty[2]

[1]Assistant Librarian, Regional Campus Library
Karnataka Veterinary, Animal and Fishery Sciences University (KVAFSU), Hebbal, Bengluru Karnataka e-mail: shantabetageri@gmail.com
[2]Compute Assistant, Regional Campus Library
KVAFSU, Hebbal, Bengaluru. Karnataka, e-mail: girish.sb3@gmail.com

ABSTRACT

This paper presents the findings of a bibliometric study one of the renowned Journal of Dairying, Foods and Home Sciences into consideration with an aim to analyse the contributions of the author and the citations cited by various articles appeared in it. The present paper covers the bibliometric analysis of year wise distribution of articles, subject wise distribution of articles and authorship of contributions. The present study analysis 34 issues containing 542 articles published in the said journal from 2003-2012. Highest number (65) of articles is published in 2012. Most of the articles 283(52%) are 1-4 pages long.

Key Word: *Bibliometric study,Journal of dairying Foods and Home Sciences, Citations,*

1. Introduction

Literature is the through content expressed by the author in published writings. It is the sensitive indicators of the emerging new ideas in any discipline. The primary role of literature is to record and transmit innovative ideas or discoveries in any specific field of knowledge that bring in advancement of knowledge and further development of a subject as well. Therefore, a careful evaluation of literature may indicate a complete picture of the discipline towards development.

The term ' Bibliometric ' was coined by Pritchard in 1969, and he defined that ' bibliometric is the application of mathematical and statistical methods for measuring quantitative and qualitative changes in collections of books, journals, and other publications'. By using quantitative analysis it is possible to measure the scattering of articles to different journals and also make it possible to measure growth and obsolescence of literature in different subjects.Bibliometrics is an emerging thrust area of research and has now become a well established part of information research and a quantitative approach to the description of documents.

The most used bibliometric methods such as citation analysis, bibliographic coupling, coward analysis and co-citation analysis can be used to map the knowledge structure and the use of literature. Bibliometric techniques have evolved over time and are continuing to do so. The counting of papers with attribution by country, by institution and by author, the counting of citations, to measure the impact of published work on the scientific community etc.

In,1948,the Father of Indian library science, Dr. S.R.Ranganathan, coined the term librametry",Which histrorically appeared first and was intended to modernize the services of librarianship. Bibliometrics is analogous to Ranganathan's librametrics, the Russian concept scientometrics, informatrics and subdisciplines like econometrics, psychometrics, sociometrics, biometrics, technometrics, chemometrics and climetrics, where methematics and statistics are applied to study and solve problems in their respective fields. Scientometrics is now used for the application of quantitative methods to the history of science and overlaps with bibliometrics to a considerable extent.

2. About the Journal

Journal of Dairying, Foods and Home Sciences(JDFHS) is an official publication of the Agricultural Research Communication Centre. This journal provides forum for the scienctific community to publish their original research articles/short communications in the field of dairying, food, nutrition, rural development, women development etc. and focuses on new developments. The journal is being published for the last thirty two years and is being scanned in the important indexing and abstracting journal of the world.

In order to examine the characteristics and trends of articles published in Journal of Dairying, Food and Home Science, a total of 542 articles from the following years were collected for analysis:

- 2003-20010 during Dr. V. D. Mudgal was the editor
- 20011-2012 during Dr. Ghanshiam Singh Rajorhia was the editor.

3. Review of Literature

Bibliometrics is the quantitative analysis of research publications. Bibliometric analysis is used in the study of scholarly communication. Vijay & Ragjavan have done citation analysis, a set of items (authors, documents, journals etc.) for the journal " Journal of Food Science and technology". Thanuskodi studies many bibliometric aspects of papers in Journal of Library Philosophy and Practice form 2005-2009 and Kulkarni, Poshett, & Narwade carried out a bibliometric analysis of the journal Indian Journal of Pharmaceutical Education and Research revealed that journals were the pre-dominant citation source followed by books. Analysis shows that majority of the scientists preferred to publish research papers in multiple authorships and there is considerable time lag in publication of articles from the date of receipt of the papers (Kulkarni, Poshett, & Narwade, 2009).. Tsay and shu conducted some statistical analysis on the citation patterns of the Journal of Documentation and found that journal articles were the most cited documents. Journals are the indicators of literature growth in any field of knowledge. They emerge as the main channel for transmitting knowledge.

4. Need for the Study

Periodicals are the indicators of literature growth in any field of knowledge. They emerge as the main channel for transmitting knowledge. Due to the escalating cost of the periodicals and lack of adequate library budgets, the selection of any particular journal for a library should be done carefully. Therefore, the library authorities are forced to reduce the number of journal subscription.

5. Objectives of the study

The principal objective of the present study is to conduct a bibliometric study of the papers reported in the Journal of Dairying food and Home science from 2003-2012 under the following objectives:

- To study year-wise distribution of papers
- To examine and study the volume wise distribution of contributions
- To examine the average number of citations per volume
- To study the subject coverage of articles
- To find out the authorship pattern of contributors
- To study the length of articles

6. Methodology

The present study Makes an attempt to analyses the bibliographic attributes of the articles published by Journal of Dairying, Foods and Home Sciences from 2003-2012: 10 Volumes (vol. 21 to 32) containing 32 issues have been taken up for the study. The information like authors, states, contributions institutions is extracted from

Journal of Dairying, Foods and Home Sciences website published by Agricultural Research communication Centre, Karnal, Haryana, India..The collected data have been analysed and is presented in the form of tables and figures.

7. Data Analysis and Interpretation

The articles published in the Journal of Dairying, Food and Home Science was manually scanned and all needed data has been recorded on a data input sheet designed for the purpose. For presenting the data in graphic the MS-Excel was used. The collected data was subjected to the analysis as per the objectives of the study.

Table 1: Year wise Distribution of articles

Year	No. of Articles	Percentage
2003	55	10
2004	42	08
2005	54	10
2006	55	10
2007	48	09
2008	50	09
2009	56	11
2010	48	09
2011	63	12
2012	65	12
Total	**536**	**100**

During the period 2003 to 2012 , 542 articles were published. Table 1 ,shows the monthly wise contributions of the journal of dairy, food and home science. 2012 year 31 volume shows the highest number of total 65 contributions. 2003 year 23 volume shows the lowest number of contributions.

Table 2: Distribution of Contributions (Volume wise)

Year	Vol. No	No. of Issues	No. of Contributions	Percentage
2003	22	3	55	10
2004	23	3	42	08
2005	24	3	54	10
2006	25	3	55	10
2007	26	3	48	09
2008	27	3	56	10
2009	28	3	56	10
2010	29	3	48	09
2011	30	4	63	12

2012	31	4	65	12
Total	**10**	**32**	**542**	**100**

The total number of contributions in the ten volume was 542. The year 2012 & 2011 shows the highest number of contributions 65(12%) & lowest contributions are in the year2004 showed the 42 (8%).

Table 3: Authorship Pattern of Contributions

No, of Authors	No. of contributions	Total No of Authorship	Percentage
1	20	20	**1**
2	201	402	**25**
3	194	582	**37**
4	77	308	**20**
5	38	190	**12**
6	12	072	**5**
Total	**542**	**1574**	**100**

Productivity pattern of authors in Journal of dairy, foods and Home science was determined by analyzing the data Table 3, workout the percentage of productivity of the authors either single or in collaborative work. On analysis of the data , majority of the papers were found to be three authored. Papers with three authors constitute the bulk of the publications (37%). Papers authored by two author constituted of (25 %) .Paper with four authors constitute the bulk of publications (20%). Papers with five authors constituted of (12%). Papers with six authors constituted of (12%). The productivity od papers with single author is negligible with the percentage of contribution is 1 %.

Table 4: Subject Distribution of Articles

Subjects	No. of articles	Percentage
Fruits and Vegetables	95	18
Food grains	125	23
Dairy Products	66	12
Dairy Farming	29	05
Animal Husbandry	35	06
Rural Development	38	07
Dairy management	30	06
Nutrient value of food	**44**	**08**
Food Package and preservation	**33**	**06**
Other	**47**	**09**
Total	**542**	**100**

Table: 5 Length Wise Articles

Pages	Year										Total	%
	2003	2004	2005	2006	2007	2008	2009	2010	2011	2012		
1-4	27	17	27	37	22	35	35	22	26	35	283	52
5-8	25	21	26	14	21	20	20	22	31	25	225	42
8-Above	03	04	01	04	05	01	01	04	06	05	34	06

Table 4 shows that a majority of contributions appeared under food grains (23%) and fruits and vegetables (18%), dairy products shows (12%), nutrient value of food shows 8 % rural development concern contributions shows 7 %. Animal husbandry, dairy management andfood package and preservation concern contributions shows 6%. In the field of dairy farming shows lowest contributions(5%).

Most of the articles 283(52%) are 1-4 pages long, followed by 225 (42%) articles with 5-8 pages, and the remaining 34 (6%) articles have the length of above 8 pages.

Table :6 Distribution of Citations (Volume wise)

Year	Vol. No	No. of Citations	Percentage
2003	22	498	09
2004	23	563	11
2005	24	416	08
2006	25	622	12
2007	26	384	07
2008	27	541	10
2009	28	543	10
2010	29	469	09
2011	30	560	10
2012	31	731	14
Total	**10**	**5327**	**100**

Table 6 indicates the range and percentage of citations per volume. The volume 31 having a maximum citations 731(14%). This is followed volume 25 having a 622 (12%) citations, volume 23(11%) citations. Volume 29 (08%) contains lowest number of citations.

Table-7 Articles and references

Particulars	No. of Contributions	Percentage
With reference	493	91
Without references	049	09
Total	**542**	**100**

Nearly all contributions have references 493 (91%) and without references 49(9%).

Table -8 Forma of documents cited

Forms of documents	Total No. of citation	Percentage
Journals	**3053**	**57**
Books	**564**	**11**
Seminar/Conference	**431**	**08**

Dissertations	965	18
Others	334	06
Total	**5327**	**100**

Majority of the articles preferred journals as the source of information which occupied the top position with the highest number of citations 3053 (57%) of the total 5327 citations. This is followed by 965 (18%) occupied by dissertations. It is followed by books 564(11%), seminar/conference 431(8%) and others 334(6%).

8. Findings

The journal has a long term history of 32 years, of which this study examined ten (2003-2012).In this long period the journal has keep up its main aim of acting as a medium for communication of all sorts of information to scientists, researchers and business mans. The present study reveals that the highest number of articles have appeared in the area of Food grains and fruits and vegetables.

- Journal published 542 articles during the period(2003-2012)
- Volume number 65 highest number of articles (12%)
- Most of the contributions are by a joint authors.
- Most of the articles 283(52%) are 1-4 pages long.
- Most of the articles are contain references 91%
- Majority of articles have appeared in the area of food grains. .

9. Conclusion

Bibliometric study of a single journal provides a portrait of the concerned journal by indicating the quality, maturity and productivity of the journal. It informs about the research orientation that the journal supports to disseminate and its influence on author's choice as a channel to communicate or retrieve information for their research needs. The majority of articles are joint authors. The study is based on 5327 citations appended to 542 articles published in 10 volumes of 32 issues appeared in Indian Journal of Dairying, Food and Home Sciences. In the year 2012 shows the maximum number of contributions made in this journal. The journal is popular among food and home science field researchers.

Reference

Das,Tappas Kumar.(2013), "A biblionetric analysis of contributions in the journal 'Library Trends', Library and Philosophy (e-journal), paper 1014

Education and Research (1996-2006) – a bibliometric analysis. *Annals of Library and Information Studies* , *56* (4), 242-248.

Kulkarni, A., Poshett, B., & Narwade, G. (2009). Indian Journal of Pharmaceutical

Kunwar P Singh, Aarti Jain, and Parveen Babbar.(2011). DESIDOC Bulletin of Information Technology: A Bibliometric Study. SRELS Journal of Information Management, Vol 48, No 1, pp.57-68.

Ranganathan, S.R."Library and its Scope" DRTC Annual Seminar 702 (1969: 285-301)

Thanuskodi S.(2010,October). Bibliometric analysis of the Journal Library Philosophy and Practice from 2005-2009. Library Philosophy and Practice, pp1-6

Tsay, Ming-yueh & Shu, Zhu-yee. (2011). Journal bibliometric analysis : a case syudy on the Journal of Documentation, Journal of Documentation,Vo 67 ,No 5, pp 806-822

Vijay, K R & Raghavan,I. (2007). Journal of Food Science and Technology: A Bibliometric study, Annals of Library and Information Studies, Vol 54, No 4, pp 207-212.

Vijayakumar P,Indian Journal of Biochemistry and biophysics (2001-2010): A Bibliometric Analysis . National Seminar on ELITE 2011 9-10 December, pp 672-678.

www.webpages.uidaho.edu/~mbolin/thanuskodi-lpp.htm

13

Digital Rights Management in Digital Libraries: An Overview

Dr.D.Ravinder

Professor & Head, Dept.of library &Information Science
S.K.University, Anatapuramu
Andrapradesh

ABSTRACT

Digital right management system aims to create a secure frame work to control access and actions that can be performed by users both human and machine. Digital rights management technologies have become very important role in increasing networked world because of its determined control over the file. Digital rights management (DRM) is emerging as a formidable new challenges and focuses on security like encryption and watermarking. It is a type of access control technology that is used by hardware manufacturers, publishers, copyright holders and individuals with the intent to limit the use of digital content. Libraries may license resources, such as images and videos, which may require a DRM system to protect the files from copying or misuse. The DRM technology is used to limit copying, printing and sharing of e-books. A properly managed DRM system could assist libraries in managing these services. The purpose of this article is to summarize the development and application of Digital Rights Management (DRM) in the digital library scenario. The author identifies the effects of DRM on the trinity of digital libraries – the creators, the content and the consumers. The paper also introduces the Open source projects on DRM. The article summarizes the technology process and effects of DRM in digital libraries within a legal frame work.

Key words: *Digital rights management, DRM tools, digital libraries, copyright, Open source software.*

1. Introduction

Information and communication technologies, internet, and particularly the World Wide Web have revolutionized the information explosion. Now the e-publishing agencies and digital libraries face the challenge to protect the authority rights and fair use of the digitized reading material. Digital rights management (DRM) is technique that attributes certain conditions on some digital products to be used and shared in libraries and information centers. The DRM is set up as a system for the protection of digital works, and created or designed to protect the unauthorized duplication and illegal distribution of copyrighted digital products. As the internet is becoming widely used, it is easy to copy and illegally sell a variety of marketed digital information and products. Therefore, this type of technique prevents users from adopting any illegal and unauthorized attempts.

The DRM systems are designed to ensure the harmony of the object so that the object is not intercepted before delivery and to ensure the security of the whole distribution chain, so that objects are transferred only to authorized consumers and devices. The DRM system makes use of the technology and tools to create an end-to-end secured packaging and distribution system for protected contents. The system generally includes the following steps:

- Watermarks and identifiers are used to identify the content uniquely. This identification can also be used for downstream tracing of the content to ensure an authorized use of the content.
- To ensure that only consumers with appropriate keys can access the content and to ensure that the content is unchanged throughout the process.
- To manage the encryption and decryption of the content by authorized entities in the content value chain.

2. Implementation of DRM Technology

For maximum utilization of the resources, the important part is its implementation in the ground level. Various techniques are used to protect the right of the author. The DRM technologies have enabled publishers to enforce access policies that not only disallow copyright infringements, but also prevent unlawful use of copyrighted works. The DRM is a copy protection tool generally used by the e-book publishers to restrict the user of converting their e-book formats from one to another and limit them to copying, printing, and sharing of e-books. The fundamental security requirement for a DRM system is that the hardware and / or software used to access the protected data be guaranteed by its manufacturer to behave in accordance with licenses. A terminal is an abstract single – user player, editor, or similar that may be implemented as a hardware device, a software application or combination of the two.

3. Tools and Components of DRM

The DRM systems usually comprise of several technologies that enable a transaction of digital documents to the authorized user. Many core technologies involve in DRM that include:

3.1 Water marking

Watermarking inserts information into content that can be used for many purposes, such as provenance (creation and ownership), copyright, and the conditions of use. Water marks are inserted into contents of a digital file that do not affect the original content. Digital information embedded within any digital media can later be detected and extracted with the water mark recognition software. Watermarks can be measured in terms of capacity and original signal fidelity.

3.1.1 Application of Water marking

A number of possible applications of watermarking technologies are there in the field of library and information science which are:

- In the field of data security, watermarks may be used for certification, authentication, and conditional access. Certification is an important issue for official documents, such as identity cards or passports.
- Another application of watermarking is on the protected identity cared or library card. The identity number is written in clear text on the card and hidden as a digital water mark in the identity photo. Therefore, switching or manipulating the photo, identity will be detected.

3.2 Encryption:

Encryption is based on cryptography. Information security is provided on computers and over the internet by a variety of methods. But the most popular forms of security rely on encryption; the process of encoding information in such a way that only the person (or computer) with the key can decode it. Encryption is a standard method to protect digital content from unauthorized used by scrambling the content, until a key is used to decrypt the content and make it usable by the key holder. The key (code) is the most important component in an encryption system. Encryption systems generally are of two categories:

a) Symmetric/Private – key encryption; and

b) Public – Key encryption (also known as asymmetric – key encryption).

3.2.1 Application of Encryption

A popular implementation of public – key encryption is the secure sockets layer (SSL) developed by Netscape. It is an Internet security protocol used by Internet browsers and Web servers to transmit sensitive information. The SSL allows authenticating both the client and the server, and also establishing a secure connection between client and server. It is primarily designed to ensure the security of electronic transactions over the internet. This SSL has become part of an overall security protocol known as transport layer security (TLS).

4. DRM Technology in Libraries

The DRM products are developed in response to the rapid increase in online piracy of commercially marketed material which proliferates through the widespread use of file exchange programs. Typically, DRM is implemented by embedding code that prevents copying, specifies a time period in which the content can be accessed or limits the number of devices the media can be installed on.

Recently, electronic document delivery services/system libraries have adopted to deliver the documents in the softcopy rather than the physical copy of the same to its user. On this system of delivery, users receive a copy of a required article which is being requested by him from the source, but this system is objectionable by the publisher of the document because when a user receives a document he is free to share it with others without any limitation, free to share it with others without any limitation. To avoid these types of misuses, libraries should have to follow the DRM technologies with the following techniques:

- Many commercially licensed resources are bundled with digital rights licenses or water marks that may be imperceptible to the libraries as well as to the end users.
- Digital signature or the hand – writing signatures are used to regulate the access to digital content, and
- E-books in the library used DRM technology to limit copying, printing, and sharing of e-books. E-books are usually limited to a certain number of reading devices and some e-publisher prevents any copying or printing. Some commentators believe that DRM makes e-book publishing complex.

5. Future of DRM

The DRM is emerging as a formidable new challenge. It is essential for DRM systems to provide interoperable services. Solutions to DRM challenges will enable untold amounts of new content to be made available in safe, open and trusted environments. The technology can be expected to be heavily used in the future to support digital library collections, software development, distance education, and networking of digital items. Libraries must not be prevented by DRM from availing themselves of their lawful rights under national copyright law and must be able to extend their services to the digital environment.

Libraries that routinely purchase DRM – protected content may want to investigate the use of modular hardware components to avoid collisions with operating systems and other application whenever a DRM component is automatically renewed. The best current strategy for dealing with embedded DRM is awareness-studying the system configuration and DRM specification that vendors provide. The DRM components of any given product are marketed to the content rights holder, not to the user.

6. DRM and Copyright (Amendment) Act 2012 in the Digital Era

'Indian Copyright (Amendment) Bill, 2012' enacted in the Parliament in May 2012, was largely based on two treaties, i.e., World Intellectual Property Organization (WIPO) Copyright Treaty, 1996 and the WIPO Performances and Phonograms Treaty (WPPT), 1996. it addresses the challenges posed by digital technology to the protection of copyright and related rights, particularly with regard to the dissemination of protected material over digital networks such as the internet and deals with copyright protection for the authors of literary and artistic works such as writing, computer programs, original data bases, musical works, audio-visual works, works of fine art, and photographs.

This Indian copyright Act includes two new Sections 65 A and 65B to punish persons found guilty of piracy by using technology to take away somebody's copyright and then use that material to make profits. So, the Copyright (Amendment) Act, 2012, makes substantial progress in filling the gaps in the parent Act (1957) so as to benefit all stakeholders. The act provides a clear picture on the rights of authors for his/her creative works.

7. DRM and the 3C of Digital Libraries:

Managing access to online information is a broad problem, which occurs in a wide range of different applications. Managers of online information wish to implement policies about who can access the information and greatly demands the awareness among users about the terms and conditions under which these implications are posses.

The Creators, Content and Consumers are the trinity of any digital library management system. Any policy framed by the managers/administrators should not affect their surveillance function, privacy and freedom of expression. This issue is based on the available technology which we are adopting in DRM system. Based on the literature and the personal experience, the authors have explained implication of DRM technology on the trinity of digital library.

8. DRM and Creator

With the DRM technology –

- The creators can have an exclusive control over their content and there need of a regulatory technology to control the access and used of general public is achieved.
- The creators have enough opportunities to promote their work, establish relationships with users and earn income by tapping into the niche markets.
- The creator will enjoy not only the authority of distribution or copying, but also the laws, contracts and licenses.
- The technology helps the creators to manage copyright material and the terms and conditions on which it is made available to users. The DRM system helps to mange creative material and protects content from copyright infringement.

9. DRM and Content

- With the available international standards the DRM technology provides option to manage and protect the interest of the copyright holders, by providing adequate identification and description tools pertaining to content availability (i.e metadata).
- The technology supports the association of the content with various applications including anti-piracy services.
- In the analog world such association between content and its metadata can be achieved by printing an identifier onto the data carrier containing the content(e.g., by printing a bar code on to a CD cover or an ISBN onto one page in a book). This approach fails in the digital world, however, because there are no physical carriers to carry the identifiers. Hence a technology is needed that allows obtaining the metadata from looking at the content itself.

10. DRM and Consumers:

- The technology imposes the restriction at the cost of the consumers' rights: to privacy, to freedom of expression, to fair use rights, and the promotion of science and the useful art.
- Copyright law gives copyright owners the right to prohibit others from making some uses of their work, such as copying, distributing or making a derivative work, where as the DRM is highly inflexible to the end user to make fair use of the works.
- Right to privacy has different dimensions. One aspect of privacy that would be directly affected by DRM is informational privacy. The administrative process of DRM demands the user personal information for authentication or registration, this has made the user to think of security of the personal information he / she submitted.

11. DRM and Open Source Movement:

Although the OSS movement objects the principles of DRM by supporting the 'Right to Read ' battle, the Open source community has contributed the tools for DRM. The requirement of such tools has discussed in the OS community for a quite a while and centers around the need for independent applications for accessing content (unprotected as well as DRM governed content) (Becker 2003). The Open Digital Rights Language (ODRL) initiative in an international effort which aims to develop and promote the open standard for rights expressions. ODRL is intended to provide flexible and interoperable mechanisms to support transparent and innovative use of digital content in publishing, distributing and consuming of digital media across all sectors and communities. The ODRL Initiative governance is managed by the ODRL International Advisory Board.

Following are the products available in the Open Source market for the DRM activities.

- OpenIPMP is an open source DRM for MPEG-4 and MPEG-2 adhers to ISO/MPEG IPMP open standards.

- OpenSDRM is an open source digital object rights management solution is another product of ODRL initiative based open-source components and on open standards.
- DRM InterOPERABILITY Frame work, an interoperable DRM architecture implementing standardized interfaces and processes for the interoperability of DRM Systems. The DRM Inter OPERAbility Framework is independent of specific hardware and operating systems and is not restricted to specific media formats. It enables user based license provision as opposed to the situation today where licenses are assigned to devices. Project Dream is an initiative to develop an open Digital Rights Management (DRM) solution for multiple domains (media, documents, enterprise, personal etc.,). This open source project develops an end – to – end Reference Implementation for the Dream Specification in order to enable a quick- start for DRM solutions.

Other OS initiatives -

- Open media platform and open DRM systems are the two software's which go long way towards solving hurdles faced by the three key elements of the digital environment community consumers, digital – media companies and consumer – electronics vendors. These hurdles include a current lack of device interoperability; media rights culture clashes among the key stakeholder; and the dearth of an easy to use secure, pervasive content solution(Foley,2003,available at
- The Pachy DRM Digital Rights Management solution is an open and suitable for any media on any device on any network. A working, cross platform code base is also immediately available to the community. The source is open and free for development (client and server), with a minimal royalty for a commercial distribution license.

12. CONCLUSION

Digital rights management system is a means of delivering content. However, DRM is frequently seen only as a technical protection measure i.e., technical means of enabling right holders to deliver digital content in a controlled way, preventing users from having access to the content unless they meet the requirements of the right holder, be it financial or otherwise, and preventing users from using the accessed content in ways other than the right holder has given permission for. Libraries are already involved in the clearance and management of rights. A property managed introduction of DRM systems in its widest sense, could assist libraries in managing their services. The challenge for the society is to balance between internet threat and the DRM policy, and to perform.

References

Choudhuri, S K.(2011). Digital right management. A Technological measure for copy right protection and its possible impacts on libraries, p.24

Harinarayana, N.S.(2009). Digital right management in digital libraries: An Introduction to Technology, effects & the availability open source Tools, 7th international caliber, p.456

http://www.defectivebydesign.org/Librarians-Aganist DRM. (accessed on 22 October 2012).

May, Christopher, (2007). Digital rights management: Problem of expanding ownership rights. Chandos Publishing, Oxford,

Paney, S (2012). Changing Mechanisms in copyright ontology: Digital Right management, p.5-6.

Retrieved from https://www.pachydrm.org/

Retrieved from (http://odl.net/).

Retrieved from (http://sourceforge.net/projects/opensdrm/).

Retrieved from (https://dream.dev.jave.net/).

Retrieved from http://jolt.law.harvard.edu/

Retrieved from http://www.extremetech.com/article2/o,3973,822282,00.asp

Retrieved from https://drm-opera.dev.java.net//

Timothy, Armstrong K (2006), Digital Rights Management and the process of fair use, Harvard Journal of law & Technology, vol20, Number 1 fall 2006,

14

Digital Agriculture Libraries Empower the Digital Agriculture

Narendra Singh Rohila[1] and B.P. Singh[2]

[1]Technical Officer, [2]Asstt. Chief Technical Officer
National Library in Dairying
National Dairy Research Institute, Karnal, Haryana
e-mail: ns_ndri@yahoo.co.in

Abstract

Right information, in right form, at right time, in right language, to the right user makes the hub of agricultural and dairy developmental cycle. Digital resources appear to be the most appropriate supporting forces of this precise, accurate and accelerating dynamic activity. Indian agriculture extension and training sector encounters different types of problems. The principal limitations are of slow operationalization of extension system and non-availability of required knowledge resources at the door step level. Since ages, libraries are playing a vital role in the dissemination of knowledge and information. Therefore, Libraries are important elements in the foundation of knowledge and advancement of rural economy. The poor resources and lack of networking library systems is also a neglected dimension of Indian knowledge infrastructure. Development in ICTs offers enough opportunities to revamp extension departments and restructure libraries to access to knowledge. Information and communication technologies now provide just in time access to global information resources. The present paper deals with a model of networking of libraries, which can provide a platform to access knowledge and its spreading far and wide to the farmers. Various issues related with compilation of databases of agricultural knowledge, library networking and dissemination to farmers will be discussed.

Keywords: *Digital Library, Digital Agriculture, ICT, ICAR,*

1. Introduction

It is true that knowledge goes on expanding if it is shared with other concerns. Libraries have stock of knowledge and information in books, journals and other reading materials, but it is in vain if it is not utilised by the society. Knowledge is useful if it is used by the stack holders for the social reform. According to Dr. S. R. Ranganathan, knowledge / books are for use and every reader has its book / knowledge. We know Knowledge / information is a mute entity, they cannot approach to its user, instead of this, users have to approach according to his / her demand. In addition to this, as per the 4th law of Ranganathan, it is also important to save the time of library user. But now time has changed, mind set up of stakeholders have also changed. They have no time to come physically in the library to access the books. With the advent to internet technology, library has been transformed from traditional library to virtual library. The different activities of libraries are now based on network technology. Now with cloud technology libraries are also available on user's mobile phones.

In India around 68% population live in villages and their basic activities are agriculture and dairy farming. The production and productivity of marginal farmers is still far behind. Therefore, it is a big challenge to increase agricultural productivity for its growing population. The reason due to wide disparities at the individual farmer's level in agriculture productivity due to lacking in percolation of information and knowledge to small and marginal farmers. There should be a mission to promote agriculture, not just to the farmers but also to every citizen and rather we must fight against agriculture illiteracy. Presently farming is not what it used to be in earlier times, it is now more of technology driven than mere labour oriented.

The important facet to enhance agriculture productivity is to ensure right information to the right farmer at the right time and in right manner. The agricultural libraries can play a vital role in disseminating the information to the farming community. The agriculture libraries must come forward and should accept the challenges and develop libraries network system of ICAR and SAUs organizations. All the libraries of agricultural and veterinary universities R & D institutions and subject matter specialists can develop knowledge bank and networks. This will lead to transfer of knowledge and skills to the farmers at their own door step or field at the appropriate time to act as a powerful catalyst to increase productivity of their crops and animals.

2. Agricultural Libraries Network

Libraries particularly those of agricultural organizations are the nerve centres of education, research, training, and extension activities for various skills of reproduction, production and disease resistance of crops and animals. The aim of digitization and networking of agricultural libraries is to improve the efficiency of accessing relevant agricultural information by the users and farmers. Agricultural libraries are changing rapidly in India and have computerized their services. There is a need to digitize the reading material for different type of users. The World Bank has funded different projects to upgrade the libraries of the institutes under the Indian Council of Agricultural Research (ICAR) and those under the State Agricultural Universities (SAUs). The ICAR has established network connectivity with all the agricultural institutes and SAUs so that the publications and information based on agriculture available in one institute can be accessed by the scientific community, policy makers, students, extension workers and the farmers of remote areas.

3. Services of Agricultural libraries

Agricultural libraries have traditionally helped to preserve the knowledge resources and serve the information needs of the contemporary society. They must quickly respond to emerging new information environment and adapt to new information handling methods and changing information search of user communities. Greater interface with professional research and ever advancing technologies is required to leverage advantage for developing better agricultural library and information systems. Usually agricultural libraries have been helping in providing information support to teaching and research programmes. Time has come when they should take a social responsibility and help in transfer of knowledge to farmers in close collaboration with extension departments. Agricultural libraries should in fact increase social interface and step into new roles wherever they find service gaps. For instance, development of database of best agricultural practices abroad that can conform to and can be customised to Indian conditions, creation of databases of Indigenous agricultural knowledge best practices, which will help in developing linkages between the innovative farmers and agricultural research institutions.

4. Next level of Agricultural libraries

The next level of agricultural services demands a dynamic role of libraries. In fact, the organization is required to take responsibility for overall knowledge management needs of the institutions where it resides. There is need to develop linkages between technologists and researchers for the creation of new knowledge. The libraries must build up interface with the society for utilization of existing knowledge and motivate the famers to use these innovative ideas at ground level.

Knowledge in itself is no power, if it is confined on the shelves of library. The repositories of a research institution as explicit knowledge or retained without use in the mind of individuals as tacit knowledge is useless for the society. The power lies in putting knowledge to use in work process. The value of library is not in keeping valuable unused collections in its holdings. Libraries should come forward to serve the society with technological advancement in the field of agriculture and also in other fields.

Indian people are presently living in digital era. Our Prime Minister Sh. Narendra Modi ji has also emphasized to transform various activities from conventional to digital. He also wants to connect rural areas with digital power. It has initiated with a project **Bharat Net** in which India's villages are to be connected via broadband network to facilitate the farmers. It is well known that right information at right time has tremendous power to change life. Mobile applications should be developed for mobile phones in facilitating the farmers access to various agricultural information such as area specific crops, fertilizers, different crop varieties, metrological information, animal husbandry, irrigation methods etc.

5. E-Resources in Agricultural Libraries

Library can play a pivotal role in dissemination of knowledge and information. Though no one library can afford to house all the documents and information on all or different subjects of agriculture. So many libraries must make a hub of specific information on various subjects in respect to agriculture and dairy sectors. A front end web portal can be developed for the end users to access the answers to specific problems in various sectors of agriculture. User friendly search options can be given through metadata. The search can be provided on different categories. At the back end some databases can be developed on different agricultural problems and answers. There should be online expert system available where farmer/user can access appropriate information. Thus agricultural libraries can develop role models of providing information for various types of queries and problems.

6. Consortia for e-Resources in Agriculture (CeRA)

With the rapid growth of information and communication technologies (ICT) and advancement of web technology, almost all reputed international journals are available on-line and can be easily accessed by researchers over the network (Fig. 1). ICAR is having network for the use of scientific community. Consortium for e-Resources in Agriculture (CeRA) has been established under the NAIP for providing online access to e-journals and resources in over 120 ICAR libraries. Efforts are going on to expand the existing information resources of ICAR Institutes/ SAUs, etc., which is to be comparable to leading institutions/ organizations universally. It is also envisaged to develop nucleate e-access culture among scientists/ teachers in ICAR Institutes / State Agricultural Universities.

Fig. 1 Functioning of CeRA

7. Krishikosh - is a digital repository of accumulated knowledge in agriculture and allied sciences, having collection of old and valuable books, old journals, thesis, research articles, popular articles, monographs, catalogues, conference proceedings, success stories, case studies, annual reports, newsletters, pamphlets, brochures, bulletins and other grey literatures spread all over the country in different ICAR Research Institutions and State Agricultural Universities (SAUs). Under the ICAR's Open Access policy, Krishikosh provides ready software platform to implement all aspects of the open access policy, similar to 'Cloud Service' for individual institution's self-managed repository with central integration. Mobile application of Krishikosh has been also developed with push SMS facility.

8. Farmers' Portals – In India millions are marginal and small farmers. Despite of this they are second to none in production and productivity. Agriculture contributes 16% of the overall GDP and accounts for employment of approximately 52% of the Indian population. A farmers' portal may also be endeavour in this direction for meeting all informational needs relating to Agriculture, Animal Husbandry and Fisheries sectors production, sale/storage of an Indian farmer. A farmer will be able to get all relevant information on specific subjects around his village/block /district or state. This information will be delivered in the form of text, SMS, email and audio/video in the language he or she understands.

Model for Networking of Agricultural Libraries

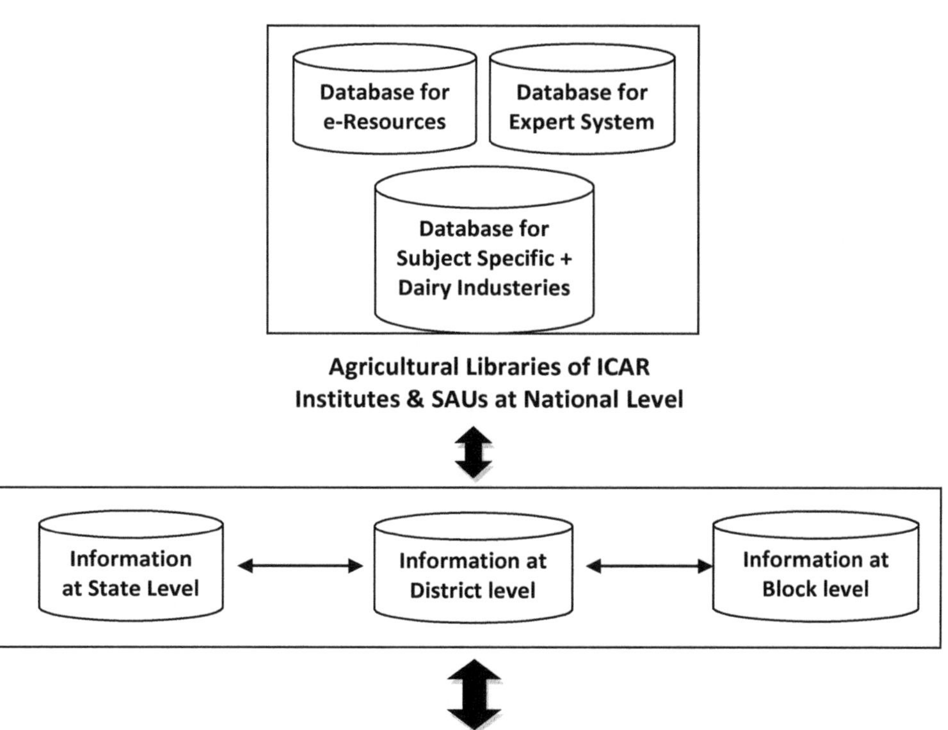

Fig.2 Information accession at Village Level

9. Audio Visual – The visual impacts more than textual. Hence, audio visual tutorials also may be developed on agriculture and dairy farming for marginal Indian farmers. In this small-small video clips can be created on various topics related to agriculture, dairy farming, dairy processing, clean milk production, use of fertilizer, soil testing etc based on area specific.

10. Working of Proposed Model

In the proposed model the repository of digitised information, databases on problems, expert system and package of practices on different agricultural dairying can be maintained at National level. In this network the farmer/user can access the related information/data at different levels such as state, district, block and village level. The farmer can get region specific solutions. A web portal can be developed where the user can access the problem specific solutions as per demand. This is a two way communication systems. The needy persons can put the query for subject matter specialist (SMS) and also can get the solutions for the problem at the same time. This is demand driven model. Therefore, this model can be effective in increasing the agricultural and dairy productivity (Fig.2)

11. Conclusion

The traditional role of libraries was to collect, organize, preserve, and disseminate intellectual outputs, however, the approaches to these activities are evolving to facilitate easier and faster access worldwide. With the onset of the digital age, the role of libraries has expanded to not only serve as a center of collected printed materials, but also to provide access to electronic information. However, libraries in developing countries are lagging and vary in their ability to provide access to knowledge resources, due to limitations in funding, power supply, Internet accessibility, infrastructure, and human capacity.

References

Chandrasekharan, H., Patle, Sarita, Pandey, P S, Mishra, A K, Jain, A K, Goyal, Shikha, Pandey, Amit, Khemchandani, Usha and Kasrija, Rajkumari (2012). CeRA - the e-journal Consortium for National Agricultural Research System; *Current Science*, 102 (06): 847-851

Jain, A. K., Chandrasekharan, H. and Kumar Rajesh (2014). Final Report of NAIP sub-project 'Digital Library and Information Management under NARS (e-GRANTH)'. Indian Agricultural Research Institute, New Delhi, India

Jain, A. K., Kumar, Amrender, Batra, Kamal, Kapur, Sanjiv (2016). Reference Manual on Krishikosh - A Repository for NARES. ICAR– Indian Agriculture Research Institute, New Delhi, pp50

Jain, Arun K., and K. Veeranjaneyulu (2013). Repository Of Indian National Agricultural Research & Education System (NARES): Open Access To Institutional Knowledge." Available at http://dspace.col.org/bitstream/handle/11599/1830

Kumar A, Sharma R, Sharma A, Batra k, Kapur S (2016). Digital Initiatives for Agricultural Research & Education under ICAR in India. pp49- 68.

NAIP (National Agricultural Innovation Project) (2014). An Initiative Towards Innovative Agriculture. Final Report, National Agricultural Innovation Project, ICAR, New Delhi

Rathinasabapathy, G., 2015. Strengthening of Digital Library and Information Management under NARS (e-Granth): An Overview: *Journal of Library, Information and Communication Technology*, *4*(1-2), pp.73-78.

Rathinasabapathy, G.; Veeranjaneyulu, K. and Amarendar Kumar (2016). Krishikosh: Institutional Repository of National Agricultural Education and Research System (NARES) of India: An Analytical Study. ICDL 2016 - Smart Future: Knowledge Trends that will change the world, New Delhi. pp. 873-887. ISBN: 9788179936535

Veeranjaneyulu, K. (2013). Knowledge Sharing among Agricultural Libraries in India: The Efforts of Indian Council of Agricultural Research. *Management Electronic Resources in Libraries, CK Ramaiah (Ed), New Delhi. India: Allied Publishers Pvt. Ltd* (2013): 75-85.

Veeranjaneyulu, K. (2014). KrishiKosh: an institutional repository of National Agricultural Research System in India. *Library Management* 35.4/5 (2014): 345-354.

15

Next Generation and Agricultural Library System

Dr. S.M.Rokade[1] and C.V.Bankar[2]

[1]Officer on Special Duty, Gondwana University, Gadchiroli
Nagpur, Maharashtra State, e-mail: rokade1134@gmail.com
[2]Asst. Chief Technical Officer (Lib & Doc)
Central Citrus Research Institute, ICAR, Nagpur, Maharashtra State
e-mail: shekhar_bankar@yahoo.co.in

ABSTRACT

The paper succinctly describes the state of art in agriculture, changing scale of information and user demands for access, informatics in agriculture, next generation realities, needs and services, products and next generation library system. It was found that the students, scientists, researchers and extension workers are the main users of library and next generation library system needs to be agile, pertinent, personalised, and convenient and compete for attention through ease of use, excellence of content and functionality space. It is concluded that in future our information services should be allow one to discover knowledge rather than to search for information and accessible from within whatever online place a user in habits and as and where networked device available. In changing scenario the libraries took place as "Knowledge Resource Centre". Therefore we need to strive for much simpler ways to find much more relevant information "one search box to rule them all" federated discovery of all available resources from a single point of entry is an overarching need.

Keywords: *Agricultural Library System, Agricultural Education, Agricultural Informatics, Information needs, Information services*

1. Introduction

The information services are keys to the development of agriculture education, research, extension education, and agriculture and agro industries in India. The services are rendered by the Agriculture University and ICAR institute libraries to their users for better development and to fulfil the aims and object of institute. In India, since 1911- 2012, during the period of hundred years the libraries moved conventional to non conventional services like microfilms to digitization, fax and telex to e-mail, land lines to mobile, cable to Wi-Fi, audio visuals, T.V., VCR, computer, and from there to laptops, and now it is the tablet in the classroom and in future the picture will be so varied.

It is now realized that in future the tremendous changes will be effected in agriculture education, research, and extension and agriculture sector and therefore it is now important to understand not only the current climate. We must also know what will be valued in the future so that we can begin to take appropriate action to improve knowledge management system to act as an efficient clearing-house of technology, knowledge and information in agriculture and allied sectors. Considering the importance of changing scenario and continuous development in library services it is necessary to know the services and products for next generation in agricultural libraries, their needs, role of future cyber agricultural knowledge resource centre, national digital agricultural information centre, for strategic thinking and planning, supporting librarians in making better decisions, improving services, saving of time and money for the future development of library and users.

2. Objectives

- To know the state of art in agriculture education, research and extension education in India
- To identify the changing scale of agricultural information and user demands for access
- To find out the need of next generation agricultural library system
- To propose integrated agricultural library system

3. State of art in agriculture education, research and extension education in India

Most people think agriculture is farming, ranching, or other forms of food, animal feed, fiber and ornamental production, processing, and consumption. However, modern agriculture encompasses much more. Agriculture and agricultural science touch every aspect of Indian society from the individual consumer's health and safety to the nation's welfare, security, and environmental sustainability.

Increasingly, agricultural research is fueling innovation in many parts of the economy not generally associated with agriculture, such as energy, electronics, plastics, and pharmaceuticals. Part of the impetus comes as agriculture continues to move into genomics, genetic transformation and other forms of plant and animal biotechnology, with applications that transcend the traditional definition of agriculture. Many major drug therapies, like Taxol for breast cancer and Artemisin, used to cure malaria, are based on plant derivatives.

Similarly, crop biomass engineering is showing promise for providing a larger share of the energy currently supplied by the petrochemical industry. Scientists also are working on ways to use biotechnology for environmental preservation and remediation using plants to detect, monitor, absorb and store toxic and hazardous substances. Other products currently under development include: proteins and enzymes for diagnostic, therapeutic and manufacturing purposes; modified fatty acids and oils for paints and manufacturing; and biopolymers as substitutes for plastics.

4. Changing scale of information and user demands for access

The exponential growth in computing, storage, and networking power along with concomitant growth in digital information and data are having a major impact on research, education, and library functions. Increasingly the Web is overtaking print as the medium for communication. Almost all scientific and technical literature is now created in digital form (termed "born digital") and large quantities have been converted to digital retrospectively (termed "reborn digital"), while new Federal guidelines require open digital publication of research results. This raises questions of how best to support the life cycle management of research and learning materials; how to develop greater systems integration among learning, library, and administrative systems; and how to integrate information skills into learning and knowledge generating activities. The traditional linear, batch processing approach to scholarly communication is changing to a process of continuous refinement as scholars write, review, annotate, and revise in near-real time across the Internet.

Research and educational materials are produced throughout universities, research labs, Agricultural Experiment Stations and Cooperative Extension offices. The diversity is immense: courseware, documents, databases, data sets, and simulations. They are hugely variable in approach, in technologies, in scope, and in complexity. Information also comes in many formats from print to digital to microfilm and audio/visual formats. Libraries expanding digital resources require new and improved tools for collaboration and for working interactively with all types of artifacts of scientific progress. This includes observed and simulated data; taxonomies; mathematical expressions; molecular, chemical and genomic expressions; structural, physical, and computational models; tables, graphs, charts, maps and images; field and laboratory notebooks; monographs and other scholarly documents; critical reviews and discourse; ontologies; bibliographic references; and remotely sensed imagery as well as sensor-generated biophysical spatial data.

As the library community works together with other organizations to develop strategies for providing efficient and effective archival, preservation, and delivery services to this plethora of information and data, user expectations have continued to expand and drive increasingly sophisticated systems for access. The majority of Indians households now have personal computers and Internet access. According to

OCLC, the graduating class grew up with computers, multimedia, the Internet and a wired world. In India the students began using computers between the ages of 5 and 8. By the time they were ages 16 to 18 all of them had begun using computers.

4.1 Informatics in Agriculture

Informatics is the study of the application of computers and statistical techniques for the management of information. It is also defined as the science of information. It is often though not exclusively, studied as a branch of computer science and information technology and is related to database, ontology and software engineering. Information is critical to social and economic activities relating to the developmental process. Indian economy has already witnessed several revolutionary developments, viz. Green (food grains), White (milk), yellow (oilseeds), Blue (fishery) and now rainbow revolution in agriculture, biotechnological revolution, industrial revolution and information technology revolution etc. The country has made considerable progress in the communication system, telephony and digital audio, video processing etc. Good communication system coupled with information technology has a great potential for providing needed support to the agriculture in achieving sustainable production by way of timely dissemination of agricultural technology needed by the farmers.

4.1.1 Observation

It is observed from the table 4.1 that the users required both conventional and non-conventional services for their use and they have emphasised on non conventional services and facilities of library like paperless, virtual, and digital with repositories with the facility of OPAC, OCLC to use anywhere, where electronic devices are available. It is found that 37.70% users expected conventional services and 99.26% user's non-conventional services in the library for their use in future.

5. User's preference

5.1 Next generation

New generation is expecting instant answers to their queries and becoming library collaborators and that library of the future should work more with people rather than for people

5.2 Products

The new technologies provide us with opportunities for customizing, developing and delivering information and knowledge in many different ways. Information managers are the best people to do this

The new products are coming on line and the users are giving preference to them due to easy and quick access and desire from the libraries simple searching without visiting to the library with the help of user ID at any where, on line availability, vast heterogeneous content, fflexibility, unification, data enhancement, integration, step-by-step evolution, open access sources, OPAC, OCLC services, and need to one search box for searching required information. In view of this

Table: 4.1 Users view towards next generation agricultural library system

<table>
<tr><td rowspan="3">Users</td><td colspan="8">Next generation agricultural library system</td></tr>
<tr><td colspan="4">Conventional</td><td colspan="4">Non conventional</td></tr>
<tr><td>Total</td><td>Recd.</td><td>Response</td><td>%</td><td>Total</td><td>Recd.</td><td>Response</td><td>%</td></tr>
<tr><td rowspan="2">Students
U.G.,P.G., Ph.D</td><td>200</td><td>180</td><td>40</td><td>22.22</td><td>200</td><td>180</td><td>178</td><td>98.89</td></tr>
<tr><td colspan="4">Print media, text books, reference books, journals, Indian & foreign, encyclopaedia, yearbook, almanacs and dictionaries in multiple languages, biographical and bibliographical sources</td><td colspan="4">e-resources, digital, audio video, repositories, OPAC, OCLC, on line direct access, net working, free access, mobile(cell) services , quick and easy services, etc.</td></tr>
<tr><td rowspan="2">Academic Staff</td><td>110</td><td>95</td><td>65</td><td>68.42</td><td>110</td><td>91</td><td>90</td><td>98.90</td></tr>
<tr><td colspan="4">reference books, journals, Indian & foreign, indexing and abstracting service</td><td colspan="4">Need of paperless, virtual library services and to use anywhere</td></tr>
<tr><td rowspan="2">Extension Workers</td><td>60</td><td>49</td><td>11</td><td>22.45</td><td>60</td><td>60</td><td>60</td><td>100</td></tr>
<tr><td colspan="4">reference books, journals, Indian & foreign, regional language publications</td><td colspan="4">One search box access, easy to use where electronic device is available</td></tr>
</table>

Source: questionnaire and survey

considering the preference of users in print and electronic collection, online searching in anywhere, quick access it is necessary to have develop integrated agricultural system on national and international level with and by effecting e-resources for the development of users.

6. Imaging Tomorrows' Agricultural Education

India is changing and so should the agricultural education. New India does not want to be trapped in the past. Today we need skills, not just degrees, with the ability to assimilate, adapt, apply and develop new technologies. Today we need high quality agricultural graduates equipped with problem solving and creative skills and ability to think and improve productivity of agricultural sector. Apart from the technical and generic skills, our graduates need leadership and entrepreneurial skills to build leading teams, and put innovations into practice and respond to competitive environments.

7. Next Generation Integrated Agricultural Library System

- Complete the migration from print to electronic collections, retire legacy print collections, redevelop library space, reposition library and information tools, resources, and expertise and migrate the focus of collections from purchasing materials to curating content
- Collect, store, organize, share, synthesize and offer computational features for huge volumes of widely disparate and distributed digital information. This includes scholarly journals, monographs, textbooks, learning objects, abstracts, manuscripts, maps, Internet resource descriptions, still images, geospatial images and other kinds of vector and numeric data, as well as moving picture and sound collections.
- Offer a federated network of institutional repositories which allows seamless, "one-stop shopping" interfaces with persistent links from a course reading list or other learning objects to the most appropriate copy of an information resource, abstracts, full-text, data, and information packages. Determine how different people need to analyze and find information and adjust to their "contexts."
- Include formats and delivery channels that enhance use of information. Provide new knowledge management tools for collaboration and for working interactively with all artifacts of scientific progress.
- There is need to undertake National Preservation Program for Agriculture Literature under ICAR, state govt and University, to identify and preserve the most significant agricultural literature.
- To increase communication (interoperability) among disciplines, there are major efforts in the library world to develop and adhere to collection standards, classification systems (schemes), common metadata structures, and other types of tagging so that eventually researchers can search, analyze and model across all platforms and formats simultaneously. Increasingly, too, academic libraries have moved from the traditional service role of

the library into project management; database design and management; interface programming; metadata.

- Integrated information system under one umbrella should be developed in the interest of users and institute.
- The creation of institutional repositories is an important first step necessary to build a national agricultural information system and "Krishiprabha" created at university library, Haryana Agricultural University, Hisar with the help of ICAR, New Delhi Generation; copyright clearance; and website hosting. Libraries are also supporting new directions in scholarly communication such as open-access publishing and self-archiving; partnerships between libraries, university presses, publishers and software developers; and the creation of institutional repositories to preserve, maintain, and provide access to institutional intellectual resources.
- The next generation library systems need to be nimble, personalized, relevant, and convenient. Our library organization needs to fully embody these traits too. The library should compete for attention through ease of use, excellence of content, and functionality of space. Our information services should allow one to discover knowledge rather than to search for information. And they should to be accessible from within whatever online place a user inhabits (Google; Face book; Blackboard), in whatever language one works in, and on whatever networked device.

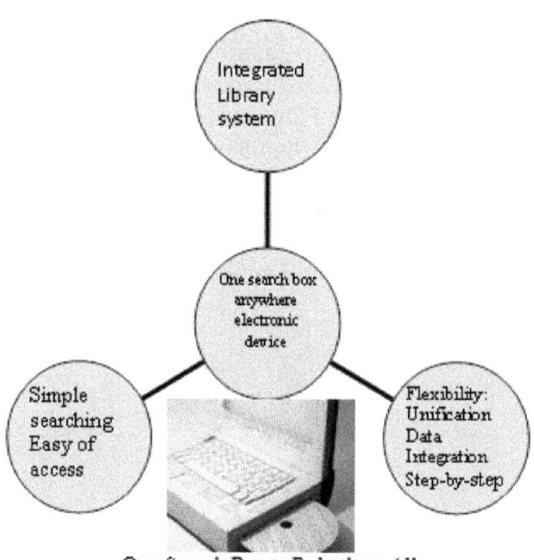

One Search Box to Rule them All

8 Conclusion

It is concluded that the next generation agricultural library systems need to be agile, quick, ongoing, relevant, personalised, convenient, and iterative. The next generation system should be defined by users for their better development and which support access, discovery, selection and filtering. The system should be easy use and to access information anywhere with the help of user ID. There is need to have a National Digital Agricultural Information System to strengthened agricultural education, research and extension education in India. The next generation service should be radically enhanced resource integration and move us on form the isolated data silos of the present.

The conventional Libraries in a technology driven knowledge society have undergone a metamorphic change in relation to the repository of information, dissemination of information, preservation of materials in different formats namely print, audio, visual and other multi-media. The effective and efficient use of ICT and internet connectivity have given new dimensions and scope for running of such knowledge centres that have now brought in new terminology for Library; namely " Agriculture Knowledge Resource Centre"

We need to strive for much simpler ways to find much more relevant information "one search box to rule them all" –federated discovery of all available resources from a single point of entry is an overarching need.

References

Barbara Hutchinson , 2006 Making the Case for a Next-Generation Digital Information System to Ensure America's Leadership in Agricultural Sciences in the 21st Century United States Agricultural Information Network (USAIN) http://usain.org/WhitePaperFinal.

Jain, Arun Kumar and Veeranjaneyulu, K (2014). Repository of Indian National Agricultural Research & Education System (NARS) : Open Access to Institutional Knowledge. Indian Journal of Agricultural Library and Information Services 30 (1), pp. 1-5.

Rathinasabapathy, G.; Veeranjaneyulu, K. and Amarendar Kumar (2016). Krishikosh : Institutional Repository of National Agricultural Education and Research System (NARES) of India : An Analytical Study. ICDL 2016 - Smart Future : Knowledge Trends that will change the world, New Delhi. pp. 873-887. ISBN: 9788179936535.

Veeranjaneyulu, K. (2013). KrishiKosh: An Institutional Repository of National Agricultural Research System in India. Library Management 35 (4&5), pp.345-354.

16

Application of Google Apps for Libraries

Dr (mrs.) Shalini R. Lihitkar[1] and Arual Joseph[2]

[1]Assistant Professor & Former Head, [2]Student MLISc –II Year
Department of Library and Information, Science
Rashtrasant Tukadoji Maharaj Nagpur University, Nagpur
Maharashtra State. e-mail: shanwaghmare@yahoo.com

ABSTRACT

Google apps is a web-based and collaborative software as a Service (SaaS) solution that customizes the proprietary Google Platform and brand for businesses of all sizes, including large enterprise. This paper attempts to profile selected Google Apps useful for libraries viz., Google Mail, Google Books, Google Calendar, Google Drive, Google Maps, Google News, YouTube, Google Plus and Google Class Room.

Keywords: *Google Apps, Google Books, Google Class Room, Library, Software as a Service (SaaS)*

1. Introduction

Google is an American multinational technology company specializing in Internet-related services and products. These include online advertising technologies, search, cloud computing, software, and hardware. Google was founded in 1996 by Larry Page and Sergey Brin in California. They incorporated Google as a privately held company on September 4, 1998. An initial public offering (IPO) took place on August 19, 2004, and Google moved to its new headquarters in Mountain View, California, nicknamed the Googleplex. Rapid growth since incorporation has triggered a chain of products, acquisitions, and partnerships beyond Google's core search engine (Google Search).

Google Apps can be defined as

- "Google Apps is a Web-based and collaborative Software as a Service (SaaS) solution that customizes the proprietary Google platform and brand for businesses of all sizes, including large enterprises.
- Google Apps facilitates the provisioning of Google applications and user/enterprise management tools, including Gmail, Google Talk, Google Calendar, Google Docs, Google Videos and Google Cloud Connect.

It offers services designed for

- work and productivity (Google Docs, Sheets and Slides), email (Gmail/Inbox),
- scheduling and time management (Google Calendar), cloud storage (Google Drive), social networking (Google+),
- instant messaging and video chat (Google Allo/Duo/Hangouts),
- language translation (Google Translate),
- mapping and turn-by-turn navigation (Google Maps)
- video sharing (YouTube),
- note taking (Google Keep),
- and photo organizing and editing (Google Photos).

Google Apps

- Google apps include everything from Gmail to YouTube and all the mobile versions.
- However, Google Apps ("Apps" with a capital "a") can also be used to refer to a specific suite of hosted services that businesses, schools, and other organizations can administer using Google's servers and their own domain.

Google Apps

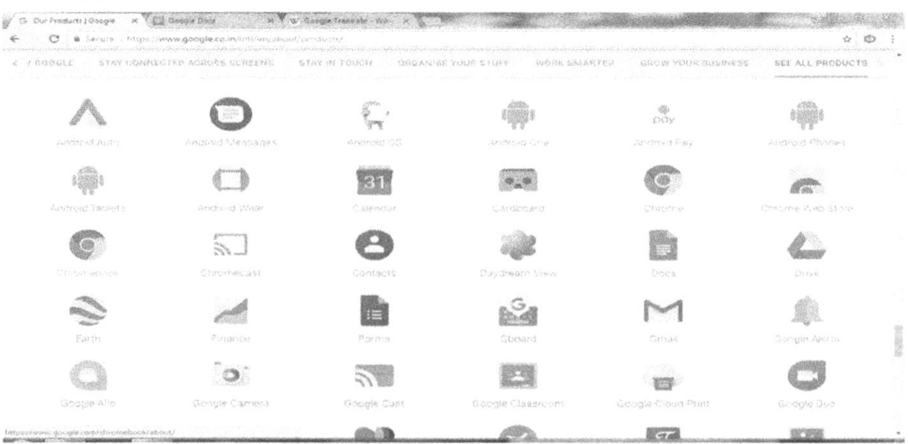

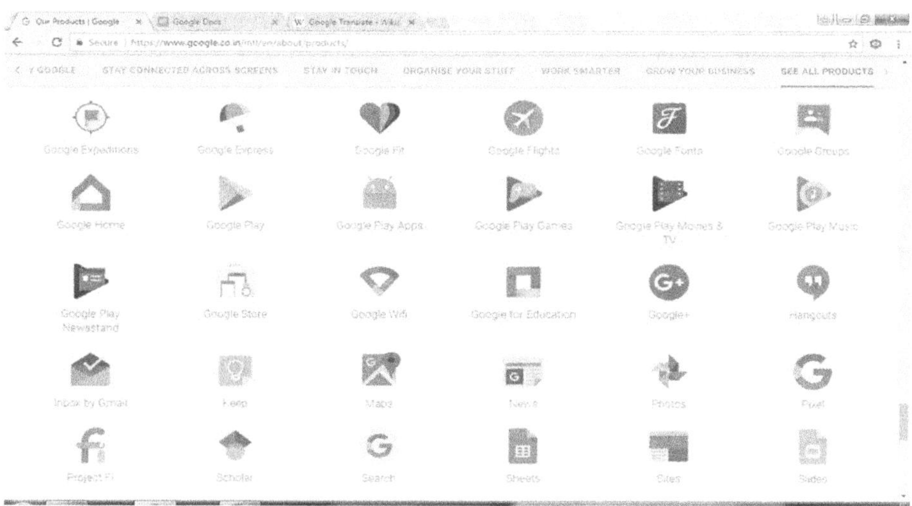

2. Selected Google Apps for Libraries

All the selected apps can be applicable to libraries for providing various services to their users. Following are the apps :

- Google Mail
- Google Books
- Google Calendar
- Google Drive
- Google Maps
- Google News
- You Tube
- Google Plus
- Google Class Room

3. Application of Google apps for libraries

3.1 GOOGLE MAIL 

- Gmail can be used in library to send and receive mails especially ordering books and finding out the details
- Using Gmail the librarian is able to send images and videos regarding library to various users.
- It is possible to have group contacts so that the information is able to reach all.
- It is possible to categorize emails especially of primary emails, educational Email and social mails.

- It is possible for a librarian using Google Mail to insert certain files from a Google drive.
- The librarian can insert photos and links about library using Google mail.

3.2 GOOGLE CALENDAR

- By using Google calendar in library the librarian is able to plan its various Activities: such as programs, seminars, lectures etc... for the whole year.
- Apart from general calendar it is possible for the library to create its own calendar.
- The library activities can be planned weekly, monthly, daily and based on agenda.

3.3 YOU TUBE

- The first thing about library regarding you tube is that it is possible for a librarian to upload small videos of orientation programmers of Library. So that the users are can have a glance of it.
- It is possible for a librarian to convert audio files into video files.
- It is not possible for anyone to upload videos rather we need to have certain formats which will suite for uploading.

3.4 GOOGLE TRANSLATER

- The librarian using Google translater in library is able to translate using text, web address, and documents.
- Today the documents come in different languages so it is guide to the librarian to translate and understand things from one languages to another.
- It is also possible for a librarian to hear the pronunciation of words using Google translate.
- It is advisable that if library has a website, it is possible to make it available in many languages simultaneously.
- It is very much useful for a librarian that if he is visiting libraries in abroad the Google Translate can help him/her to understand the meaning of the words.[1]

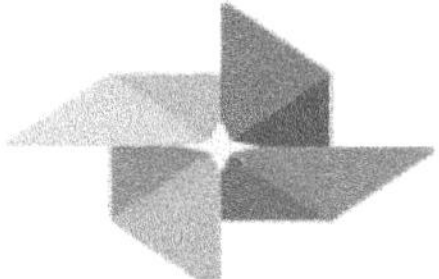

3.5 GOOGLE PHOTOS

- In library there are collection of photos of various activities, programs etc. so all those can be uploaded in Google Photos. And these can be viewed later when it will be required.

3.6 GOOGLE BOOK

- Google book helps librarian in library to search desired books online means.
- The librarian can advise the users the availability of books regarding various fields.
- The librarian can write his own review about certain books using Google Books.
- The librarian can add books from Google books to his own library so that the users can benefit reading in the library itself

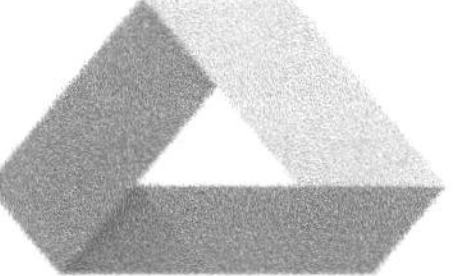

3.7 GOOGLE DRIVE

- The Google drive has a storage of minimum 15GB. But this can vary from device to device. So the librarian can store some of the library materials in Google Drive by channeling into different folders: especially library files and documents etc.

3.8 GOOGLE NEWS

- Apart from the basic activities in library, the Google News helps the librarian and the users to have a knowledge up to date because it provides news about sports, technology, science etc.
- So Google News keeps the librarian and the library up to date.

3.9 GOOGLE MAP

- The library should be registered its location in internet world. So that anybody wants to locate the library can use Google Map and access the library facilities.

4. General use of the apps from the point of a librarian

Apart from the above individual applications of the apps, there are common way of application of these apps in library. They are:

1. Librarian are not just teaching face to face staff development classes, but are serving as role models by sharing document, collaborating and using these tools as teachers.
2. Librarians can be the daily "go-to" support for students and staff members for one on one help with the apps including using documents for file storage, sharing documents for viewing and editing and solving log on? Access/ password problems.
3. Librarians can teach students these apps as part of the information and technology curriculum. Using Google docs for activities they would have used desktop programs for in the past. Librarians must accept major responsibility for helping students use these resources safely and responsibly.
4. Librarians can create templates and tools to help student during research process. Google Apps is less cumbersome than managing separate wikis, blogs, and other for project- long documentation since everyone's work is in the same place and easy to organize.
5. Librarians can use the Apps for their own library information gathering efforts- conducting surveys, teaching classes via shared calendars and organizing data.
6. Librarians can use e-mail groups and chat for communication with their staff and students, making information available in real time – not relying just on monthly newsletters.
7. Librarians can use Google documents for curriculum writing, lesson planning, and collaboration with teachers and with other members of library department. Librarians can use and model self-made video tutorials shared Google video.[2]

5. Findings

a) First and foremost the user must have internet facility to operate Google Apps for desktop and mobile version.
b) Secondly by having one Google account the user can operate all the Google Apps.
c) It is possible to interconnect these Apps and can bring out new result.
d) It is not with one attempt one can have mastery over the Apps rather constant usage is required.

e) The user needs to have technical mind then there can be lot of discoveries and new findings both in the Google world and in the library when it is applied.

6. Suggestions

f) It will be better if the user has some computer basic knowledge then he/ she can slowly expand ones knowledge.

g) Library should run short-term courses, organize training programs using Google Apps for students and faculty members.

h) These Google Apps can be used for promoting library services.

i) Instruction should be given by libraries for users about Google Apps.

7. Conclusion

'Google it' means, look up something using the extremely popular Google search site. So in a way Google Apps means search something using the Apps to discover something. As mentioned in the introduction, with technical mind all these apps can be utilized and all these apps can be applied in the library and its activities and both librarian and users can benefit library facilities using Google apps. As now in this IT era,users are more techno savvy and they may be searching all the time on internet. So Libraries should attract our users in virtual world by using Google Apps for providing ready reference services to student as well as for faculty members.

References

Anne Rutledge, Patrice. My Google Apps. USA: Que Publishing Copyrite, 2017.

Darbyshire, Paul. Getting Started with Google Apps. USA: friendsof,an A press Company, 2010.

Google Apps: Introduction, accessed on 7th march, 2017.

Hahn, Jim. The Best 100 free Apps for libraries. UK: The Scarecrow Press. Inc. Plymouth, 2013.

http://dailygenius.com/7-great-things-can-google-classroom/. Accessed on 29th May, 2017.

http://dictionary.cambridge.org/dictionary/english/app. Accessed on 16 March, 2017.

http://panmore.com/google-vision-statement-mission-statement. Accessed on 28 February, 2017.

http://www.care2services.com/care2blog/9-youtube-features-you-may-not-know-about. Accessed on 31st May, 2017.

http://www0.sun.ac.za/ctl/wp-content/uploads/2013/08/Google-Apps-Introduction_Step-by-Step.pdf. Accessed on 3 March, 2017.

https://betterhumans.coach.me/5-useful-features-of-google-calendar-d6739989384d. Accessed on 2nd June, 2017.

https://en.wikipedia.org/wiki/G_Suite. Accessed on 25 February, 2017.

https://www.digitalunite.com/guides/social-networking-blogs/google-plus/what-google-plus. Accessed on 2nd June, 2017.

https://www.google.com/maps/d/viewer?mid. Accessed on 31st May, 2017.

https://www.laptopmag.com/articles/google-plus-improvements. Accessed on 30th May, 2017.

https://www.techopedia.com/definition/26622/google-apps. Accessed on 27 February, 2017.

Johnson, Doug. The Indispensable Librarian: Surviving & Thriving in School Libraries iInformation Age. England: An Imprint of ABC, 2010.

Miller, Michael. Using Google Apps. USA: Que Publishing Copyrite, 2008.

Primary Research Group, Primary Research Group Staff: Medical & Other Scientific Librarian Use of Google and Its Features & Apps.

Roy, Lachlan. Essential Guide to Google Apps, accessed on 7th march, 2017.

techcrunch.com/2009/06/18/google-books-adds-new-features-and-tools/. Accessed on 1st June, 2017.

Teeter, Ryan. Google Apps for Dummis. NJ: Wiley Pubishing, Inc, 2008.

VaibhavManohar and Shalini R. Lihitkar, Web 2.0 In Libraries, New Delhi: Studera Press, 2017.

www.lifewire.com/google-accounts-apps-differences-1616381. Accessed on 2nd June, 2017.

17

Webometric Analysis of English Newspapers' Websites IN INDIA

Sonia Bansal

Assistant Librarian
Guru Angad Dev Veterinary and Animal Sciences University
Ludhiana, Punjab
e-mail: soniapta@gmail.com

ABSTRACT

The aim of this study is to evaluate English newspapers' websites by using Alexa Internet that facilitate access to commercial web traffic data and analytics.Five highest circulated English language newspapers included in this study were selected from language wise list of highest circulated newspapers compiled by Audit Bureau of Circulation.Data pertaining to bounce rate, time on site, page views, search traffic and audience geography were collected from Alexa Internet website.

Keywords: *Newspapers, Webometrics, Alexa Internet*

1. Introduction

Newspapers are great source of latest and up to date information. With the advancement of ICT, most of the newspapers have started their websites/online editions to disseminate latest news to the readers. Majority of the newspapers have web presence these days.Therefore, it becomes imperative to evaluate newspapers' websites by using webometric methods to increase the usage of websites. In this study, Alexa Internet tool is used for evaluation of newspapers' websites. Alexa Internet, Inc. is a California based company that provides access to commercial web traffic data and analytics.It was founded as an independent co. in 1996 and was acquired by Amazon in 1999.Alexa's traffic ranks are based on browsing behavior of people and traffic data provided by users in Alexa's global data panel over the past 3 months. It provides global/Indian rankings, traffic data, audience geography and other information.

2. Objectives

This study aims to evaluatethe websites of English newspapers using Alexa Internet that provides access to commercial web traffic data and analytics.

3. Methodology

For this study, 5 highest circulated English language newspapers were selected from the language wiselist of highest circulated newspaperscompiled by Audit Bureau of Circulation. The websites of each of these five newspapers were searched in Alexa Internet website to collect data for bounce rate, time on site, page views, search traffic.

4. Analysis and Discussion

4.1 Popularity

Table 1 shows that The Telegraph newspaper occupies first rank in Alexa Internet global rank. The Hindu occupies second place and Hindustan Times is at third place.

Table 1 Newspapers' Websites Global and Indian Rank

Name of the newspaper	Global Rank	Indian Rank
The Times of India	222	13
The Hindu	1265	100
Hindustan Times	1100	111
The Telegraph	22361	1841
The Economic Times	222	13

In case of Indian rank, The Telegraph occupies first rank, followed by Hindustan times and The Hindu newspapers.

4.2 Bounce Rate

This figure shows that Hindustan Times has highest bounce rate (75.30%) followed by The Telegraph (66.40%) and The Hindu (62.80%).

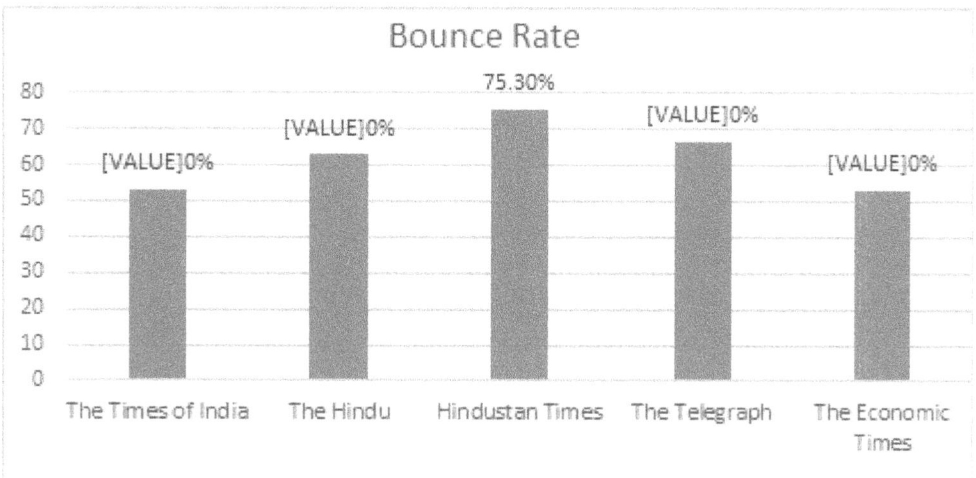

Figure 1 Newspapers' Websites Bounce Rate

4.3 Time on Site

The Times of India and The Economic Times has highest time spent on site by the visitors per day and Hindustan Times has lowest time spent on site by the visitors per day.

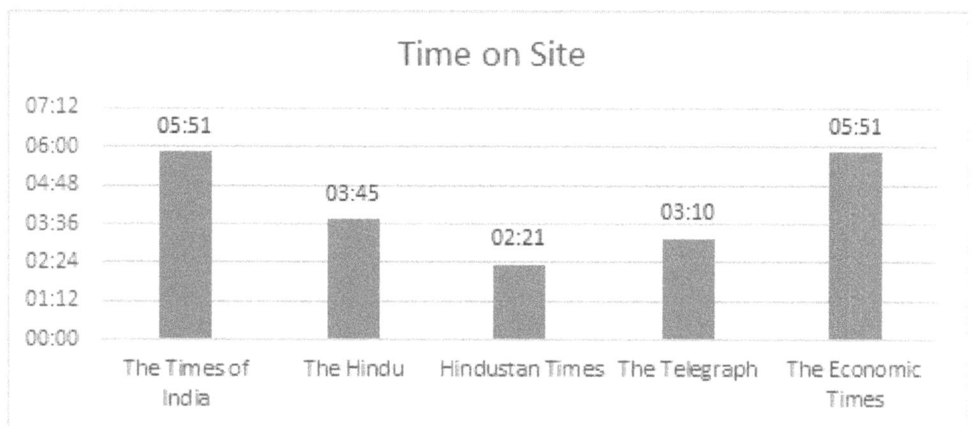

Figure 2 Newspapers' Websites Time on Site

4.4 Page views

The Telegraph has highest no. of page views by visitors daily (2.95) followed by The Economic Times and The Times of India (2.90%). Hindustan Times has lowest no. of page views by visitors.

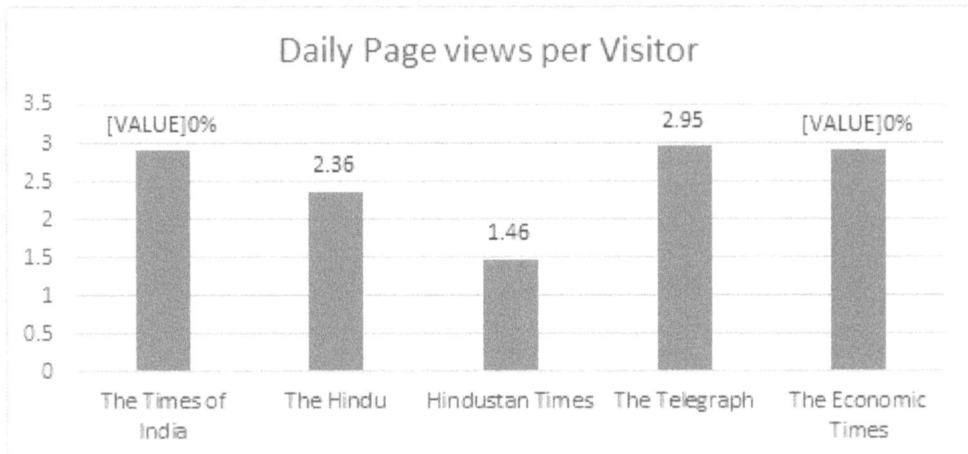

Figure 3 Newspapers' Websites Page Views

4.5 Search Traffic

Figure 4 shows that Hindustan Times holds 1st place in search traffic (36.80%) followed by The Telegraph (35.40%) and The Hindu (26.00%).

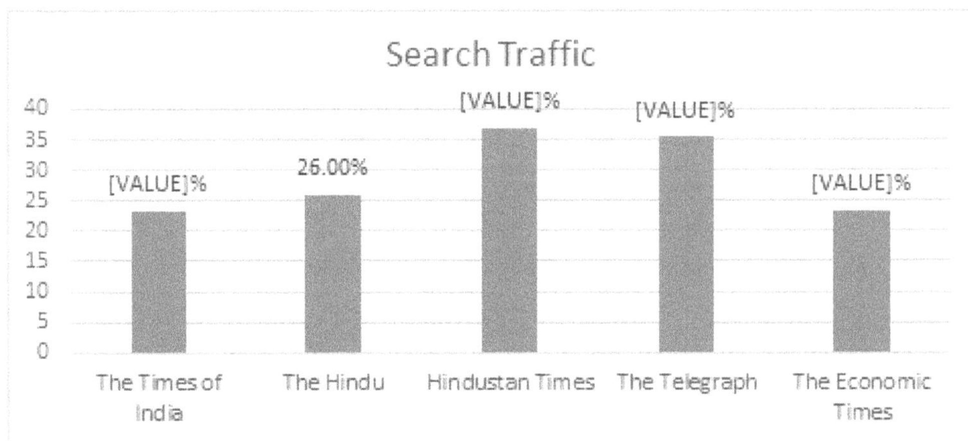

Figure 4 Newspapers' Websites Search Traffic

4.6 Audience Geography

Hindustan Times has highest percentage of foreign visitors (31.7%) followed by The Times of India and The Economic Times (26.1%) each. The Telegraph newspaper website has lowest percentage of foreign visitors (18.7%).

Table 2 Visitors by Country

Name of the newspaper	**Indian Visitors**	**Foreign Visitors by country**								
		US	UK	Pakistan	UAE	Qatar	Saudi Arabia	Bangladesh	Oman	Singapore
The Times of India	73.9%	7.8%	1.4%	1.4%	1.2%	-	-	-	-	-
The Hindu	79.8%	6.0%	1.1%	-	-	1.7%	1.3%	-	-	-
Hindustan Times	68.3%	6.6%	-	5.8%	-	-	1.5%	1.6%	-	-
The Telegraph	81.3%	4.9%	1.1%	-	-	-	-	-	1.0%	1.1%
The Economic Times	73.9%	7.8%	1.4%	1.4%	1.2%	-	-	-	-	-

5. Conclusion

This study brought forward that the website of the Telegraph newspaper is most popular and has global rank of 22361. Hindustan Times newspaper website has highest bounce rate (75.30%), search traffic (36.80%) and foreign visitors (31.7%). The findings of this study will be helpful for website managers of newspapers to increase the usage of their websites.

References

Alexa Internet.http://www.alexa.com/

Audit Bureau of Circulation. http://www.auditbureau.org/files/Highest%20Circulated%20 amongst %20ABC%20Member%20Publications%20(language%20wise).pdf

How are Alexa's traffic rankings determined?https://support.alexa.com/hc/en-us/ articles/200449744-How-are-Alexa-s-traffic-rankings-determined-

18

A study on the use of Consortium for Electronic Resources in Agriculture (CeRA) by Post Graduate Research Scholars of Punjab Agricultural University, Ludhiana

Amit Kumar

Assistant Librarian
Punjab Agricultural University
Ludhiana, Punjab
e-mail: 14ak25@gmail.com

ABSTRACT

The digital information resources available online are increasing at an exponential rate, several practices have evolved for the economic and effective delivery of such information to the end users. In this context, consortia-based information services have gathered momentum world over during the last few years. Though, there are several library consortia in India, UGC Infonet is mainly meant for universities controlled by UGC and CeRA is meant for agricultural universities. This paper discusses utilisation of consortia-based digital information resources by the post graduate of the Punjab

Agricultural University, Ludhiana. The study has been made among post graduate students to identify the trend of using CeRA.. Again the purpose of use, strategy of developing their affiliation to the consortia, usefulness of CeRA, receipt of document and attitude towards the consortia have been dealt.

Keywords: *Consortia, CeRA, e-Journals, Agriculture, Users' Study, Internet,University library*

1. Introduction

The advent of electronic or digital journals, online journals, magazines, etc., have necessitated a thorough redefining of the serials control system of the libraries as online magazines or journals are disseminated through internet websites. The information and communication technologies have made radical changes in the conventional system of information processing and its delivery and most of the universities offer courses in their multiple campuses and also through their affiliated or recognised colleges/institutions. In many universities, courses are extensively being offered through distance education mode also. Although, it is a great challenge to ensure effective coordination and communication, several programmes have already started in India to provide a seamless supply of information resources by way of access to journals and databases, course materials, online lectures, etc., in electronic format to the academic and research community. The UGC, ICAR, AICTE, National Knowledge Commission, National Knowledge Network, National Innovation Council, etc., have done leading roles towards this end. As the universities in India are broadly controlled by two agencies like UGC and ICAR, the library and information resources management and services are also performed differently. While the universities under UGC have formed UGC Infonet for consortia mode of journal subscription and electronic information services, the agricultural universities and agricultural research institutes formed CeRA for the same purpose. Moreover, there are several other consortia also in operation in India for institutions under CSIR, DRDO, IITs, IIMs, etc.

Punjab Agricultural University Library was established with a meager collection of 200 books in 1959 in College of Agriculture. However, at present this temple of learning with its beautiful five storey building with covered area of 93,320 sq. ft., centrally air-conditioned and surrounded by lush green lawns, dotted with beautiful ornamental trees and pollution free environment has grown into one of the best libraries of the region – a place of pilgrimage for scholars and faculty members from all over the country. This library has 760 seating capacity in its Five Reading Halls. The aim of the library is to provide rich knowledge to its users and to extend it further through information technology, automation and networking.

Active Members as on 31th May, 2017: 3749

Library collection as on 31th May, 2017

1	Books	254164
2	Theses	38177

3	Bound Periodicals	104502
4	E-documents (Books and Theses)	4367
TOTAL		401210

2. Objectives of Study

The purpose of the study was to explore the awareness and usage of consortia-based information and knowledge resources by the post graduate (PG) students of PAU. The main objectives were to:

i i) To study about the awareness of CeRA among the students.

ii ii) To find the purpose of using the CeRA.

iii iii) To study the use of CeRA.

iv iv) To study the satisfaction level of the users.

v vi) To study the problems faced by the students in accessing CeRA.

3. Methodology

A questionnaire was designed to gather primary data which was distributed among 110 post graduate students. Out of which only 90 responses were received. The method used for data collection was based on Random Sampling technique. The investigator also had discussions with some users on various issues pertaining to e-resources to make the data and information more convincing and authentic.

4. Scope and Limitation

The study is based on the user behaviour and the pattern of the utilisation of information resources by the PG students of PAU. The scope of the study is limited to the digital resources mainly available through CeRA and in PAU.

5. Review of Literature

Review of related literature is an essential part of social science research. A researcher is required to scrutinize the research findings of similar works, if any, done previously. Researchers

of Library and Information Science have taken up a number of research projects on user studies for all types of libraries. Since there is a tremendous growth in research activities in LIS, it is essentially required to make a review of literature to determine the quantum of work done in general and to identify research works in subjects of interest in particular.

Though a large number of user studies on library and information services of users of individual disciplines have been conducted at various levels by the LIS researchers but, no comprehensive and comparative study seem to have been made exclusively for agriculture and its related disciplines together to discover a distinct difference in their need and approach to information. The investigator has made a comprehensive search of literature of published work that are given as under:

Brennan et.al. (2002) in their study at the University of Illinois, Chicago revealed that all the faculty members, read electronic journals at least weekly and many used them on daily basis. They stated that the users of e-journals accepted them because they see more advantages of e-journals including time saving, ease of use and powerful searching capabilities.

In 2004, Tenpoir et.al. done a study on usage of e-journal by medical faculty of the University of Tennessee Health Science Center (UTHSC), reported that usage of electronic journals gradually becoming important. The electronic journals available to faculties have increased the average reading. The main purpose of the accessing electronic journals is to support the research work. In comparison with usage of print journals medical faculty still continue to rely on the print journals instead of electronic journals. Almost five years later, continue with their study Tenopir et.al. (2009) pointed out how faculty members in science, technology, medicine and social sciences from 1977 to the present in a university, read and used scholarly articles and how this reading pattern has changed with the pervasive accessibility of electronic journals. They found that the average number of readings per year amongst science faculty members continues to increase while the average time spent per reading is decreasing.

In the Indian context a study conducted by Mounissamy et.al. (2005) on research scholars and faculty members of National Institute of Technology, Tiruchirappalli reveals that 33% of faculty members and 67% of researchers frequently access the electronic journals for different purposes. 53.4% of the respondents access 1-3 journals in a month. 26.1% access 3-6 electronic journals in a month. 12.5% access 6-10 electronic journals in a month and 8% access above 10 journals.

Madhusudhan (2008) in his study on Use of UGC-Infonet ejournals by research scholars and students of the University of Delhi, Delhi', opined that it appears to be some need for academics to be provided with training in using e-journals. The result showed that most of the (72%) respondents are accessing ejournals through department computer laboratories, 47% accessed from central library. 73% of respondents agreed that print journals are important for the foreseeable future. 64% of the respondents strongly attest to the fact that they need proper training/ orientation for searching e-journals. 75% respondents stated that they are accessing UGC Infonet e-journals for current and up-to-date information.

Patil et.al. (2009) in their survey on faculty members and research scholars in Gulbarga University found that 72.01% researchers and 75.50% faculty members search information from the electronic journals. 66.83% of respondents used both the electronic and printed journals. 64.20% of respondents stated that they feel the need of more number of e-journals in UGCInfoNet consortium. Need of training and orientation programmes was emphasized by 59.19% of respondents to access e-journals.

As per the study of Singh and Tiwari (2010) all the agricultural university libraries are connected by networking system. While the under graduates are primarily using networks for communication purpose through social networking system group, the post graduate students are using for dissertations and other research assignments. Teachers are using Internet for support of their research work and preparing papers for journals. It is experienced that author and title searches are conducted on the web using Google and Google scholar.

Singh et al (1996, pp.9-13) in their study on, "electronic journals on library and information science", traced out that, electronic journals clearly have potential advantages in terms of flexibility over printed journals, however, can diminish these perceptions. Mahesh and Ghosh (1998, pp.67-76) in their study on , "availability and use of indigenous database by S & T libraries: a case study" pointed out that, science and technology libraries in Delhi are willing to buy and make use of indigenous databases compared to traditional resources.

6. Analysis

6.1 Age wise distribution of the respondents

Table 1 shows age wise distribution of the post graduate students

Age in years	PG Students	Percentage
21-25	78	86.67%
26-30	12	13.33%
31-35	0	0
Total	90	100%

Table 1 above shows that 86.67% of the respondents belong to the age group of 21-25 years, followed by 13.33% lying in the age group of 26-30 years.

6.2 Frequency of visit to library and use of CeRA

Table 2 depicts the frequency of visit to library for use of CeRA by the post graduate students

Table no. 2

Frequency	PG Students	Percentage
Daily	11	12 .22%
Twice a week	18	20.00%
Weekly	22	24.44%
Fortnightly	15	16.67%
Once a month	08	8.89%
When Needed	16	17.78%
Total	90	100%

From the above table no 2 it is clear that 12.22% of the PG students use CeRA daily. 20% of use them twice a week. Majority of them 24.44% use it weekly. 16.67% use these e-resources fornightly and only 8.89% use them once a month and 17.78% of them use them as and when required by them.

6.3 Awareness about the availability of CeRA

Table 3 shows the awareness about the Consoritum Table no. 2

Awareness	PG Students	Percentage
YES	85	94.44%
No	5	5.56%
Total	90	100%

The above table shows that 94.44% of the students are aware about the availability of the Consortium based e-resources and only 5.56% of them are not aware about them.

6.4 Opinion about sources providing information about Consortium

Different sources providing information about CeRA e-resources are

Table no. 4

Sources	PG Students	Percentage
Library Professionals	24	26.67%
Teachers	28	31.11%
Internet	15	16.66%
Colleagues	17	18.89%
Any Other	06	6.67%
Total	90	100%

It is quite clear from the table no. 4 that the information about CeRA e-resources is available through many sources. The library professionals were able to provide information to 26.67% of the users. While 31.11% of their respective teachers share the information regarding these e-resources. 16.66% of the students are able to get this information from the internet. Only 18.89% of them consult their colleagues regarding the CeRA e-resources. A small number i.e. 6.67% of them get information from other sources like department notice boards, social networking sites etc.

6.5 Place of Accessing the CeRA

Table 5 shows the various places of accessing e-journals

Table no. 5

Places	PG Students	Percentage
MSR Library	50	55.56%
Department	22	24.44%
School of Information Technology	13	14.44%

Any Other	05	5.56%
Total	90	100%

The above table shows that more than 55.56% of the post graduate students consult the M.S.Randhawa Library for accessing the CeRA e-resources. While 24.44% of them access CeRA in their respective departments. Further 14.44% of them use the computer lab of School of Information Technology and only 5.56% of them use other places like student's home, their respective hostels etc. using their own laptops and other gadgets.

6.6 Access points for searching the information from e-resources

Table no.6 shows the various access points for searching the information

Table no. 6

Searched By	PG Students	Percentage
Author	42	46.67%
Title	23	25.56%
Subject Heading	13	14.44%
Keywords	12	13.33%
Total	90	100%

It can be seen from the above table 6 that the users have different approaches for searching the information from CeRa. 46.67% use author approach for searching, 25.56% search with title approach. 14.44% like searching with the help of subject headings and 13.33% search with the help of key words. The search option using author is most popular among the post graduate students.

6.7 Purpose of Access and Use of CeRA

Table no.7 shows the purpose of access and use of CeRA consortium

Table no. 7

Purpose	PG Students	Percentage
Full text Article	35	38.89%
Abstract	15	16.67%
Request for Article	22	24.44%
Article Received	18	20.00%
Total	90	100%

The table no. 7 shows that 38.89% of the students accessed and used CerA full test resources and 16.67% used abstracts only. 24.44% respondents requested articles and 20% received articles through the Document Delivery system of CeRA.

6.8 Purpose of using CeRA

Table no. 8 shows the purpose of using CeRA

Table No.8

Purpose	PG Students	Percentage
Writing Papers	25	27.78%
Projects	09	10.00%
Notes	08	8.89%
Research Work	34	37.78%
Seminars	12	13.33%
Any Other	02	2.22%
Total	90	100%

It is quite evident from the table no.8 that 27.78% of the students use these resources for writing papers. Though 10% of them use them for project work. 8.89% use them for preparing their notes. Majority of them i.e. 37.78% use them for their research work. Further 13.33% use them for preparing their seminars and only 2.22% use them for other purposes.

6.9 Opinion regarding Infrastructure facilities for e-resources

Table no.9 shows the satisfaction level regarding infrastructure facilities

Table no. 9

Frequency	PG Students	Percentage
Fully satisfied	40	44.45%
Partially satisfied	37	41.11%
Unsatisfied	13	14.44%
Total	90	100%

The above tabulated data clearly indicates that 44.45% of post graduate students are fully satisfied with infrastructural facilities available to them in the library. However, 41.11% of them are partially satisfied with the infrastructure and only 14.44% are not satisfied with the e-resources facility.

6.10 Opinion about problems faced in using CeRA

Table 10 shows the problems faced in using CeRA

Problems	PG Students	Percentage
Lack of Training	27	30.00%
Ease of using	13	14.44%
Slow speed	32	35.56%

Abstracting & Statistcs not available	07	7.78%
Hardware and Software	11	12.22%
Total	90	100%

The data in above table no.12 reveals that 30.00 % of students felt the need of proper training for using the CeRA e-resources. 14.44% find it difficult to use these resources for their respective purposes. Slow speed of accessing the information is also the concern of 35.56% of the post graduate students. 7.78% of them, feel that abstracting and statistics are not available. 12.22% of the students feel that he latest hardware and software is also one of the requisites of providing flawless accessibility to CeRA e-resources.

6.11 Opinion about attending training for using E-resources

Users have found the use of electronic resources more convenient to obtain desired information. However, without any training programme it is not possible to obtain the information quickly. Students can improve the use of electronic resources if they are provided suitable computer training programme. The data shown in Table 11 represent the distribution of PG students of PAU interested in attending any computer training programme to improve the use of electronic resources.

Table 11

Interested in attending Training Programme	PG Students	Percentage
YES	84	93.33%
No	06	6.67%
Total	**90**	**100%**

The data in above Table no.14 shows that 93.33% of the post graduate students are willing to undergo training in improving their skills for accessing e-resources and only 6.67% of them are not interested in attending any type of training.

7. Findings of the study

The study arrived on the following findings based on the results from the analysis of the data gathered:

i The post graduate students are fully aware about the availability of the CeRA in the library.

ii The M.S.Randhawa Library was found to be the clear choice for the accesibility of CeRA.

iii The Library professionals and Internet were revealed as the chief source for information about CeRA.

iv Majority of the students are aware about the e-journals consortium for i.e. CeRA.

v The study has revealed that majority of the post graduate students access the e-journals using author approach. Further most of them are satisfied with the infrastructure available to them. The slow speed of the internet has been the main issue for accessing the CeRA.

vi The majority of the post graduate students use CeRA for their research work.

vii The study further shows that majority of the respondents are willing to undergo training for effective use of CeRA.

8. Suggestions

Based on the results and opinions of the respondents, the present study suggests the following:

i To provide the e-resources service effectively and efficiently, more number of access terminals should be installed in the M.S.Randhawa Library and other departments.

ii To save the precious time of the users high speed internet connection should be provided.

iii Need for more trained and skilled staff, having awareness of the functioning of the both software and hardware, who can help the users in areas like accessing, downloading and proper utilization of e-journal resources.

iv In order to improve the efficiency of the users towards access to electronic resources, the library should provide hand on experience and conduct user orientation programmes for the library users at the start of every academic session.

v Proper feedback system should be introduced to know about the various problems faced by the research scholars to solve them effectively.

vi More number of the e-journals should be subscribed by the library in the fields of specialization, where only small number of e-journals are being subscribed.

vii An adequate number of the research scholars suggested that necessary arrangements should be made to access the full text of more e-journals. The document delivery service should provide articles requested speedily.

9. Conclusion

CeRA of ICAR has become a heavily-used service by the students. Curriculum-based information literacy courses like 'library and information services', 'research methodology', etc., have contributed much towards imparting required skills for the access and use of digital information resources. Strengthening of CeRA services by adding more resources and facilities will provide strong information support for the education, research, and extension programmes. The present study sought to examine the use of CeRA by the post graduate students and the result shows that most of the objectives have been met satisfactorily. Using of the CeRA for research purpose in their area of the study has been one of the most important aspect of the study. Unorganized, lack of training, speed to access have been major problems faced by the students, whereas most of them are satisfied with the facilities available to them. However, majority of them feel that the user training will be of great use for increasing the usage of CeRA among the users in the library.

References

Bostick . "Sharon L (2001). Academic library consortia in the United States: An Introduction". Library Quarterly 11.(1):6-13.

Brennan, M.J., Hurd, J.M., Blecic, Deborah, D.B. & Weller, A.C. (2002). A snapshot of early adopters of e-journals: Challenges to the library. *College and Research Libraries*, 63(6), 515-526.

CeRA-Consortium of e-Resources in Agriculture. Available at http://cera.jccc.in/about/aboutCeRA.pdf

Francis, A.T. (2012): "Evaluation of Use of e-Resources in Agriculture in context of Kerala Agricultural University." DESIDOC journal of Library and Information Technology. 32.(1):38-44.

Koovakki, D.Z. & Noor, K.V.H. (2006). Electronic information use among the faculty. *Library Herald*, 44 (4), 313-320.

Lancaster, F.W. (1995). The evaluation of electronic publishing. *Library Trends*, 43(4), 518-527.

Lebowitz, G. (1997): "Library services to distant students: An equity issue". J. Acad. Librarianship,23.(4):302-08.

Madhusudhan, M. (2008). Use of UGC-Infonet e-journals by research scholars and students of the University of Delhi, Delhi: A study. *Library Hi Tech*, 26(3), 369-386.

Mahapatra, Rabindra K.(2013). "Use of Consortium of E-resources in Agriculture(CeRA) in the Central Library of Odisha University of Agriculture and Technology: A study". Asian Journal of Library and Information Science.5. (3-4): 39-47.

Mounissamy, P. & Rani, S. (2005). Evaluation of usage and usability of electronic journals. *SRELS Journals of Information Management*, 41(2), 189-205.

Nabi, Hasan . (2012): "Web-based Agricultural Information Systems and Services underNational Agricultural Research System".DESIDOC Journal of Library & Information Technology, 32.(1): 24-30.

Patil, D.B. & Parameshwar, S. (2009). Use of electronic resources by the faculty members and research scholars in Gulbarga University, Gulbarga: A survey. *SRELS Journal of Information Management*, 46(1), 51-60.

Punjab Agriculture University. Available at www.pau.edu

Selvaraj, A. D., and G. Rathinasabapathy (2016). A Study on Electronic Information Use Pattern of Faculty Members of Self-Financing Engineering Colleges in Tiruvallur District, Tamil Nadu. Asian Journal of Library and Information Science 6.3-4 (2016): 31-40.

Selvaraj, A. D., and G. Rathinasabapathy. (2015). Information Use Pattern by Students of Self-Financing Engineering Colleges in Tiruvallur District, Tamil Nadu: A Study. Journal of Library, Information and Communication Technology 6.3-4 (2015): 45-54.

Srivastava, Ranjana. (2002): "Information use pattern of researchers in chemistry: A citation Study". IASLIC Bulletin, 47.(4) :185-92.

Tenopir, C., King, D.W., Edwards, S. & Wu, L. (2009). Electronic journals and changes in scholarly article seeking and reading patterns. *Aslib Proceedings*, 61(1), 5-32.

Veeranjaneyulu, K. and Ravi Kumar, N.P.(2015). Consortium for e-Resources in Agriculture (CeRA) : A great gateway to e-Journals. Indian Journal of Information, Library & Society 28 (1-2), pp. 110-116.

19

Masive Open Online Courses (Moocs) In Library And Information Science: Need Of The Hour

Dr. Vaishali P. Gudadhe (Choukhande)

Professor and Head
Department of Library and Information Science,
Coordinator, Women Studies Centre,
Sant Gadge Baba Amravati University, Amravati
vgudadhe020@gmail.com

Abstract

An attempt has been made to explain the concept of MOOCs. MOOCs has been introduced in India since 2016, but very few are aware about the online courses. The teachers also should create awareness and take initiative to encourage the students to enhance their knowledge with the advance courses, through online mode. SWAYAM has not introduced Library and Information Science courses yet. LIS Education also needs to be updated with the advance technology and courses.

Keywords: *MOOCs, Distance Education, Higher Education, LIS Education,*

1. Introduction

A massive open online course (MOOC) is an online course aimed at unlimited participation and open access via the web. In addition to traditional course materials such as filmed lectures, readings, and problem sets, many MOOCs provide interactive user forums to support community interactions among students, professors, and teaching assistants (TAs). MOOCs are a recent and widely researched development in distance education which were first introduced in 2006 and emerged as a popular mode of learning in 2012. (Laura, 2014).

2. MOOCs and Open Education:

In April of 2001, the president of MIT, Charles Vest, announced the establishment of a project for placing MIT course contents on the Web for free access by anyone with an Internet connection. In effect, this announcement started the OpenCourseWare (OCW) movement. Since that time, open educational resources (OER) and more recently massive open online courses (MOOCs) have proliferated. A flurry of research reports, books, programs, announcements, debates, and conferences related to MOOCs and open education have encouraged educators to reflect on how these new forms of educational delivery might enhance or even transform education.

In France, OpenClassrooms, which had been offering online tutorials in IT and programming languages since 1999, begins producing MOOCs in 2012. It now offers more than 1000 courses, focusing on technology and digital skills, mainly in French, but also in English and Spanish, produced either in house, or in partnership with universities and companies.

According to *The New York Times*, 2012 became "the year of the MOOC" as several well-financed providers, associated with top universities, emerged, including Coursera, Udacity, and edX. (Smith 2012). In all 24 Providers were found providing Open Education platform, out of which 17 found commercial and 7 were non-profit. USA ranks first in providing Open Education to the learners in 14 Universities/Institutions, followed by Ireland, Australia, Germany, Western pacific region and India. Since 2007 WiziQ is the provider and IIT Delhi is one of the participant University for the Online Courses.

3. MOOC's and Higher Education:

Specialist organisations such as Coursera and edX provided the information technology (IT) platforms to support MOOCs, because managing electronic communication with hundreds of thousands of learners worldwide was beyond the IT capabilities of all but the largest open universities. Today MOOC platforms are available in countries across the world. The UK's FutureLearn consortium, for example, has 80 partner universities and institutes, which offer hundreds of MOOCs to over 3 million people. MOOCs provide a good example of our tendency to overestimate the significance of innovations in the short term whilst underestimating their long-term impact. The early predictions of a revolution in higher education proved false, and the idea that MOOCs could be the answer to the capacity problems of universities in the developing world was especially silly. Nevertheless, MOOCs are a significant phenomenon. Over 4,000 MOOCs are available worldwide and register 35 million learners at any given time.

4. MOOC's and Higher Education in India:

India is the second biggest market for MOOCs (massive open online courses) in the world, following the US. In time, however, India may surpass the US. After all, India's population is second to China's and India is third in terms of university enrolment worldwide; respectively the US and China are first and second for university enrolment at the moment but this may soon change.

MOOCs represent a huge opportunity for Indians in terms of an open education revolution. It could potentially give millions access and availability to high quality learning if they have Internet connectivity. First, there are more applicants than slots at top Indian universities. Second, millions of Indians live in poverty and are unable to afford or gain access to a higher education.

For these reasons and many others, US-based MOOC providers are entering partnerships with Indian colleges or universities or making alternative arrangements to make MOOCs more accessible and available in India. The Indian Institute of Technology Bombay (IIT Bombay), for example, is the first college in India to join not-for-profit edX and to offer MOOCs. The partnership was created to fill a specific need in India: training engineering teachers. The partnership will extend an open engineering education to a global audience, though the US and India is the largest populations of edX learners worldwide and will be the primary recipients of the partnership of an open engineering education.

Coursera and Udacity (both for-profit ventures) have also been making arrangements to make MOOCs more accessible and available in India. Coursera is working on a mobile application so students from poorer backgrounds can access MOOCs on Akash tablets. It is also offering a course on web intelligence and big data with the Indian Institute of Technology Delhi (IIT Delhi). Perhaps not surprisingly, since Coursera currently produces the largest number of MOOCs worldwide, they have seen a huge increase in Indian student enrolment as well.

Udacity, too, has seen Indian enrolment increase over the past year. In May, Udacity teamed with Georgia Tech and AT&T to offer the first online massive open online master's degree program in computer science for less than $7000 in tuition. The program is specifically targeted to India and the Middle East. The deal is seen as revolutionary within higher education circles and many administrators are eagerly awaiting the results.

Table 1 Coursera Enrolees

S.N.	Countries	Course Enrolees
1	United States	27.7%
2	India	8.8%
3	Brazil	5.1%
4	United Kingdom	4.4%

5	Spain	4.0%
6	Canada	3.6%
7	Australia	2.3%
8	Russia	2.2%
9	Rest of world	41.9%

5. MOOC's and UGC Initiatives

Under the 'Digital India' Initiative of Government of India, one of the thrust areas is 'Massive Online Open Courses (MOOCs)'. Ministry of Human Resource Development, Government of India has embarked on a major initiative called 'Study Webs of Active Learning for Young Aspiring Minds' (SWAYAM), to provide an integrated platform and portal for online courses, covering all higher education, High School and skill sector courses. SWAYAM is an indigenous (Made in India) IT Platform for hosting the Massive Open Online Courses (MOOCs).

The journey of SWAYAM can be traced back to 2003 with the initiation of the NPTEL (National Programme on Technology Enhanced Learning), a joint programme of IITs and IISc. This was the first major attempt in E-learning in the country through online Web and Video courses in Engineering, Science and humanities streams. The UGC has provided the guidelines to the universities for development and implementation of massive online open Courses (MOOCs) for necessary action. It is said that MOOCs have been tested to a degree and would be a huge investment in India›s higher education as well as contribute to one of the biggest open education revolutions in the world right now. SWAYAM has not introduced Library and Information Science courses yet.

6. Library Oriented MOOCs Courses

With the help of MOOCs courses, librarians can gain hands-on experience with particular technologies that can help improve their reference and instructional services delivery. There is a broad range of course offerings in areas of computer science (specifically, learning new coding languages) for librarians with an interest in developing technological expertise. There are also several business and management courses that may satisfy the need for librarians in managerial positions who desire a more theoretical approach to human resources management, organizational behavior, or strategic thinking. There are so many library related courses including management and Leadership, Marketing, Technological skills, job application skills etc. MOOCs LIS courses are as following:

1. The University of North Carolina, Chapel Hill is offering course on "Metadata : Organizing and Discovering Information" on Coursera platform which is only eight weeks since it is launched.
2. University of London offers course on Research Methods through the Coursera platform which also may be useful for LIS professionals.
3. School of Information Studies, Syracuse University is offering MOOCs in New Librarianship.
4. Harvard offers a 12 weeks course on "copyright" on edX platform which can be useful to library professionals

5. University of Toronto offers Mooc on "Library Advocacy Unshushed: Values, evidence and action"

Figure 1

7. Conclusion

Several universities like Stanford, Berkeley, MIT, Harvard are offering Moocs in different subjects. Apart from these EdX, Coursera, Udacity, Moocs University, OpenEd are some of the platforms that offer Moocs in different subjects from different universities.

The role of a teacher is also transformed. Teacher involved in offering Moocs have to perform multiple roles of – lecturer, designer, mentor, institutional marketer and so on. The teacher has to be more creative and have good skills of technological tools.

Students still are apprehensive about MOOCs because of the credibility, lack of accreditation, lack of peer group interaction, certification which they feel as a reward to their learning. Prime minister of India launched MOOCs platform in India on 15th August 2016, 23.6% of Indian population is below poverty, who cannot afford paid online courses. If the university or colleges take the initiatives to share or provide financial support to the students, it will benefit them to take advance courses for their betterment. LIS education should impart such advance online courses for their students through online mode.

References

MOOCs Direcotory. Available at http://www.moocs.co/Higher_Education_MOOCs.html. (Accessed on 29 June 2017)

Pappano, Laura. (2017) The Year of the MOOC. The New York Times. Retrieved June 2017.

Smith, Lindsey (2012) «5 education providers offering MOOCs now or in the future». 31 July 2012

20

Use Of Cloud Computing Technology In Libraries: An Overview

Syed Fayyaz Mohsin

Senior Research Associate,
INFORMATICS, Drug Discovery Research
Wockhardt Research Centre
Aurangabad, Maharashtra State
e-mail: fsyed@wockhardt.com

ABSTRACT

Now a day's Technology is growing fast Cloud computing is a new technology model for IT services which many businesses and institutes are adopting. It avoids locally hosting multiple servers and equipment and constantly dealing with hardware failures, software installations, up-gradations and compatibility problems. For many organizations, cloud computing can simplify processes and we can save time and money by using cloud computing. Cloud computing technology is the fifth generation of computing and the biggest thing since the Web came up as a boon for library systems and is offering various opportunities for libraries to connect their various services with clouds. There is potential for a lot of confusion surrounding the definition of cloud computing. This paper mainly presents an overview of cloud computing, cloud service models and its burning applications that can be clubbed with library services on the web based environment.

Keywords: *Cloud computing, Library, Information Technology, ICT*

1. Introduction

The explosion of the cloud computing over the past few years has led to a state to invent many new technologies. Cloud computing is a method of computing in which dynamically scalable and a lot of virtualized resources are provided as a service over the Internet. Cloud computing is performing computing tasks via a network connection while remaining is isolated from the complex computing hardware and networking infrastructure. Cloud computing can be defined as "A model for enabling ubiquitous, convenient, on demand network access to a shared pool of configurable computing resources that can be rapidly provisioned and released with minimal management effort or service provider interaction" . Cloud computing to put it simply means ***Internet computing*** i.e. computation done through the Internet. With cloud computing users can access database resources via the internet from anywhere, for as long as they need, without worrying about any maintenance or management of actual resources.

Today we are living in age of Digital Information. We can use Cloud computing in library section for collection of e –books, storage of data base organization processing and analysis of information and retrieval. In the field of higher education it has become one of the strongest adopters of virtualization as it allows the organization of all resources like laboratories and libraries centrally and gives remote access to students through mobiles too. Today's Libraries and IT experts are facing new challenges in managing electronic content achieves. The adoption of technology that enables an organization to understand the meaning of every piece of information to ensure quick and appropriate access when needed is required there are a variety of cloud-based services in the library world. Nowadays libraries are using cloud computing technology for enhancing the services by adding more values, attracting the users and cost effectiveness. In the cloud computing environment, clouds are vast resource pools with on demand resource allocation and a collection of networked features. The need of cloud computing may occur due to the information explosion, problems in accessing the information, save the time of the users and staff, resource sharing problems, problems in library resources management, complex demand of users and attraction of users towards cutting edge technologies.

2. What is Cloud Computing?

Cloud computing is not a new technology that suddenly appeared on the web but it is a new form of fifth generation of computing. Cloud computing is a kind of computing technology which facilitates in sharing the resources and services over the internet rather than having these services and resources on local servers/ nodes or personal devices.

Cloud computing is a way of providing various services on virtual machines allocated on top of a large physical machine pool which resides in the cloud. Cloud computing comes into focus only when we think about what IT has always wanted – a way to increase capacity or add different capabilities to the current setting on the fly without investing in new infrastructure, training new personnel or licensing new software.

In cloud computing, the word *"Cloud"* is used as a metaphor for the Internet, based on the standardized use of a cloud-like shape to denote a network on telephony schematics. The "Cloud" element of Cloud Computing can be seen as an acronym that stands for:

C-- Computing resources,

L-- that is Location independent,

O-- Can be accessed via online means,

U-- Used as an Utility &

D-- is available on Demand.

According to Wikipedia, cloud computing is the delivery of computing as a service rather than a product, whereby shared resources, software and information are provided to computers and other devices as a utility (like the electricity grid) over a network (typically the Internet).

According to Ellyssa Kroski, Library Journal (09/10/2009) Cloud Computing means using the Web services for our computing needs which could include using software applications, storing data, accessing computing power, or using a platform to build applications.

According to Mladen A. Vouk, "Cloud Computing is a service oriented architecture, reduced information technology overhead for the end user, greater flexibility, reduced total cost of ownership, on demand services and many other thing" Michael Armbrust (et.al) defined 'cloud computing refers to both the applications delivered as services over the Internet and the hardware and systems software in the data centres that provide those services.

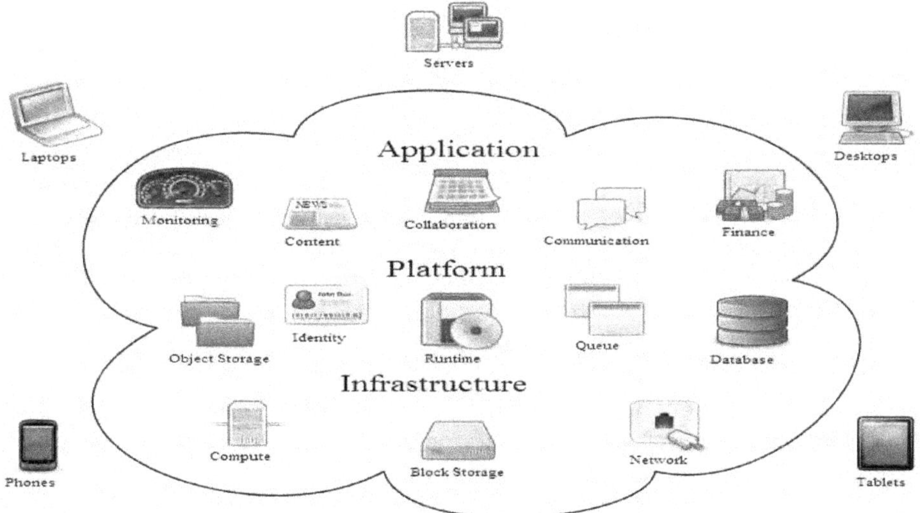

3. Evolution of Cloud Computing

Cloud computing has evolved from the most emerged technologies like grid computing, virtualization, utility computing in distributed computation environment with web based platforms. The concept of Cloud Computing came into existence in the year 1950 with implementation of mainframe computers, accessible via thin/static clients. The cloud computing has evolved from the concepts of grid, utility and SaaS.

The development towards cloud computing started in the late 1980s with the concept of grid computing. Grid computing also termed as On Demand Computing focuses on moving a workload to the location of the needed computing resources, which are mostly remote and are readily available for use. A grid is a cluster of servers where huge task could be divided into smaller tasks which will be run in parallel systems. From this point of view, a grid could actually be viewed as just one virtual server and require applications to conform to the grid software interfaces. In the 1990s, the concept of virtualization was expanded beyond virtual servers to higher levels of abstraction. Storage and network resources, and subsequently the virtual application, which has no specific underlying infrastructure were applied in virtual platform. Utility Computing is a concept established by John McCarthy, who predicted already in the late 1960s that "computation may someday be organized as a public utility". In utility computing, clusters are presented as virtual platforms for computing with a metered business model. Characteristics of clusters are that the computers being linked to each other are normally distributed locally, and have the same kind of hardware and operating system. Therefore cluster work stations are connected together and can possibly be used as a super computer. The utility approach also known as pay-per-use or metered services increasingly common in enterprise computing and is sometimes used for the consumer market for Internet service, file sharing, web site access and other applications.

More recently software as a service (SaaS) has raised the level of virtualization to the application, with a business model of charging not by the resources consumed but by the value of the application to subscribers. In 2001, IBM started autonomic computing also called self correction in which computers can automatically correct themselves without human intervention. For example, consider a network of computers running a set of programs and when there is a hardware failure on one of the computers on the network, the programs running on that computer are transferred to other computers in the network. The following section discusses the great features exist with cloud computing which made an end user to use this computing concept easily.

4. Characteristics of Cloud Computing

The essential characteristics of the cloud computing model were defined by the National Institute of Standards and technology (NIST) and have since been redefined by number of architects and experts. According to NIST, Cloud computing is a model for enabling ubiquitous, convenient, on-demand network access to a shared pool of configurable computing resources (e.g., networks, servers, storage, applications, and services) that can be rapidly provisioned and released with minimal management effort or service provider interaction. This cloud model is composed of five essential characteristics, three service models, and four deployment models.

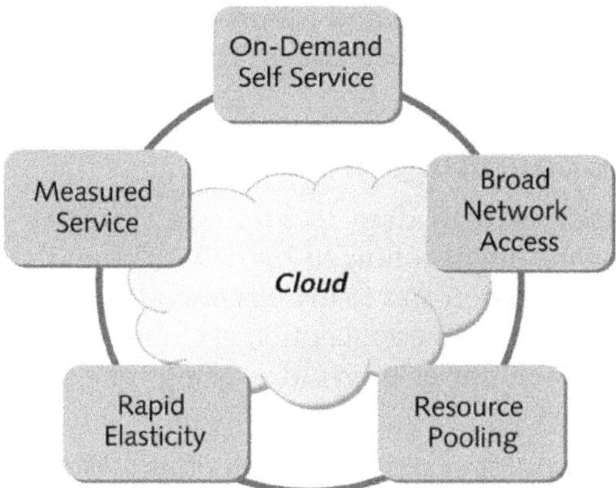

Figure 1.2 The essential characteristics of cloud computing

1. ***On-demand self-service.*** A consumer can unilaterally provision computing capabilities, such as server time and network storage, as needed automatically without requiring human interaction with each service provider.
2. ***Broad network access.*** Capabilities are available over the network and accessed through standard mechanisms that promote use by heterogeneous thin or thick client platforms (e.g., mobile phones, tablets, laptops, and workstations).
3. ***Resource pooling.*** The provider's computing resources are pooled to serve multiple consumers using a multi-tenant model, with different physical and virtual resources dynamically assigned and reassigned according to consumer demand. There is a sense of location independence in that the customer generally has no control or knowledge over the exact location of the provided resources but may be able to specify location at a higher level of abstraction (e.g., country, state, or datacenter). Examples of resources include storage, processing, memory, and network bandwidth.
4. ***Rapid elasticity.*** Capabilities can be elastically provisioned and released, in some cases automatically, to scale rapidly outward and inward commensurate with demand. To the consumer, the capabilities available for provisioning often appear to be unlimited and can be appropriated in any quantity at any time.
5. ***Measured service.*** Cloud systems automatically control and optimize resource use by leveraging a metering capability at some level of abstraction appropriate to the type of service (e.g., storage, processing, bandwidth, and

active user accounts). Resource usage can be monitored, controlled, and reported, providing transparency for both the provider and consumer of the utilized service.

Cloud Computing offers great benefits for organization and individuals by reducing cost and increasing flexibility. A good and robust Cloud Computing model should consist of all these five essential characteristics. There are also privacy and security concerns.

5. Use of Cloud Computing in Library

With the recent advancement in data technology, libraries became machine-controlled with the advancement followed by networks and virtual Libraries .To increase the ability of cooperation and to make a major, unified presence on the net the library community will apply the conception of cloud computing. This approach to computing will facilitate libraries to avoid wasting time and cash whereas alter workflows.

- Most library computer systems are built on pre-Web technology
- Systems distributed across the Net using pre-Web technology are harder and more costly to integrate
- Libraries store and maintain a lot of constant knowledge lots of and thousands of times
- With library data scatter across distributed systems the library's Web presence is weakened
- With libraries running independent systems collaboration between libraries is made difficult and expensive
- Information seekers work in common Web environments and distributed systems make it difficult to get the library into their workflow
- Many systems are only used to 10% of their capacity. Combining systems into a cloud environment reduces the carbon footprints, making libraries greener these improvements can be grouped into three basic areas: technology, data and community. Each offers some general and some unique opportunities for libraries.

6. E-Library Services

E- Library refers to all the library resources that are available online through computers and databases. This is different from the open internets because E-Libraries have restricted access. E-library system facilitates library operations by offering

- ✓ Systematic records of the library collection
- ✓ Reliable records of library patrons
- ✓ Picture perfect check out and check in of library materials
- ✓ Ease of accessing statistical results

- ✓ Generate real time report for management decision
- ✓ Outstanding/Overdue loan
- ✓ Periodic loan transactions
- ✓ Bother free stock taking of library materials
- ✓ Personalized service to each patron

Specific access account to library patrons to search or make reservation of library materials anywhere and anytime at their conveniences.

7. Advantages of Cloud Computing in Library Services

Following are the advantages of using cloud computing

- ❖ Service oriented architecture: the cloud is provided which has access to resources, software, networks, and applications through web, which is controlled by remotely located data centres.
- ❖ Pay per use model: it works on demand. We can demand the service for certain period like for few days or few weeks.
- ❖ Cost effective: The resources, services, software etc are shared by group of institutions by cutting down the individual institutes cost. Comparing to the traditional method of computing, cloud computing billing may be comparatively less.
- ❖ Portability: since the service is available over the web, the service can be availed through browser from any part of the world.
- ❖ Eco-friendly: since it is pay for use model, consumption of electricity will be minimum. Hence, it helps green computing.
- ❖ Adjustable storage: in the traditional system, if the server is less than what we have. The server should be replaced with the new one. In this computing, the storage capacity can be adjusted according to the needs of the institute, since the storage is controlled by the service provider.
- ❖ Flexible and Innovative: new technologies will be informed as and when available with the service provider and the service utilized will be more flexible when comparing with the traditional computing.
- ❖ Cloud OPAC: Most of the institutes in the world are having the catalogue over the web. These catalogues are available with their institutes local server made it available over the web. If the catalogue of the institutes made it available through cloud, it will be more benefit to the users to find out the availability of materials.
- ❖ When the data comes to cloud, the data becomes cloud, which can be shared among the users. The need for storage in local server, installation, maintenance and backup is removed so that the librarians can concentrate on innovative services.

8. Limitations of Cloud Computing in Library Services

Any technology will have its own limitations. Below mentioned are some of the limitations in Cloud Computing.

- ❖ Moving to the trusted cloud computing service will become a challenging task.
- ❖ Security: Library deals with information and has large volume of information. In order to have cloud computing the data has to be uploaded to the cloud machine. Hence, there should be strict service level agreement before entering into the process.
- ❖ Reliability: reliability is the big question in cloud computing. Once entered in to cloud computing, if the companies satisfy as per the service agreements, it will be good. Otherwise there is a chance of having discomfort.
- ❖ Data back up; intellectual property rights are the other problem which has to be taken care before.

9. Conclusion

Concluding it can be said that cloud computing technology provides libraries an opportunity to improve their services and relevance in today's information society. It can bring several benefits for libraries and give them a different future. It helps libraries to deliver its resources, services and expertise at the point of need, within user workflows and in a manner that users want and understand. It should free libraries from managing technology so they can focus on collection building, improved services and innovation. Library automation is a concept of data collection, data storage and data retrieval and we can use cloud computing effectively in library automation as it is cost effective, flexible, innovative, open access to access data anywhere, anytime as per users need. It can be used through internet. This approach to computing can help libraries to save time and money while simplifying workflows.

References

Davies, K. The "C" Word. *BioIT World*, November/December 2009, pp. 24-26, 42.

Gregory Chockler, Eliezer Dekel, Joseph JaJa, Jimmy Lin,"Special Issue of the Journal of Parallel and Distributed Computing: Cloud Computing",Journal of Parallel and Distributed Computing, Volume 69,Page 813,2009.

Lewis G. (2009). Cloud Computing: Finding the Silver Lining, Not the Silver Bullet. Available at http://www.sei.cmu.edu/newsitems/cloudcomputing.cfm

Luo, Lili (2013): Reference Librarians' Adoption of Cloud Computing Technologies: An Exploratory Study, Internet Reference Services, 17(3/4), 147-166

Matt Goldner, "Winds of Change: Libraries and cloud computing", OCLC, (2010)

http://www.oclc.org/content/dam/oclc/events/2011/files/IFLA-winds-of-change-paper.pdf

Michael Armbrust (et.al), "A view of cloud computing", Communication of the ACM, 53, No. 4, (April 2010): 50-58.

Robert L. Grossman, "Compute and storage clouds using wide area high performance networks", Future Generation Computer Systems, Volume 25, Pages 179-183, 2009.

Wikipedia .Cloud Computing. Available at http://en.wikipedia.org/wiki/CloudComputing

21

Consortia Projects in Agricultural University Libraries: AN OVERVIEW

Datta Y. Gawli[1] Gajanan P.Khiste[2] and D.B. Maske[3]

Professional Assistant[1], Information Scientist[2]
Dr.Babasaheb Ambedkar Marathwada University, Aurangabad
e-mail: dattagawli84@gmail.com
Librarian[3], Shri Panditguru Pardikar Arts College
Sirsala, Tq.Parli-Vaijinath, Dist.Beed
e-mail: dbm2289@gmail.com

ABSTRACT

In this article given information about consortia, Need of Consortia, Objectives of the consortia, Selection Process of Consortia, Movement of Consortia based Approach, Consortia Projects in Agricultural University Libraries.

Keyword *– Consortia, Agriculture, University Library*

1. Introduction

With advent of ICT technologies, in the phase of information society in the internet era the rapid changes have been taken up in the form of information and presentation. The internet has revolutionized the collection, organization, dissemination and preservation of information which have become a main function of any university library in agricultural field. In case of use of electronic journals, the publications of the documents are undergoing numerous changes, which are in Digitized form available on online as well as offline. The online publications are specially linked with the journals. The major advantage of electronic journals especially the online are as under

i The time lag between the publication and availability of the journals get Minimal.

ii It becomes easy both for the publisher and the user to manage the matter.

iii The manipulated use of the journal at a time by many people can be possible

iv Barrier of the use of the journal in library building vanishes.

v Rapid revision; and

vi Fast contact between the generator of the literature and the user.

No research can be completed without journals, hence journals seems to be important documents in university libraries. 50 to 60 percent of the library budget is generally spent on online journals as well as hard copy of print journals. The term electronic journal can be used for different type of serials available in electronic format inclusive of journals, magazines, newspapers, newsletters, etc. e-journals also incorporate publishers and journal homepages and pre-print servers. During the late nineteen ninety the use of e-journals increased in USA and the other part of the world. Various commercial services are available for the libraries to make use of the online journals. The table mentioned below is significantly in this perspective. Therefore, university libraries have also adopted the e-journal form especially in Research and Development Libraries. University libraries are also started their initiatives is using e-journals. Infonet Project of UGC-INFLIBNET is a major boon of this. eShodhSindhu is also making a role as a special consortium in the e-journals than the existing efforts of consortia like FORSA of DAE, INDEST and others, especially because of subject coverage. Consortium of electronic Resources in Agriculture (CeRA) is the glaring example of consortia activities. Therefore, it is worthwhile to give brief information on consortia based activities in India in general.

2. Consortia

A consortium is an association of two individuals companies, organizations or governments (or any combination of these entities) with the objective of participating in a common activity or pooling their resources for achieving a common goal.

3. Need of Consortia

a) Information Explosion

b) Multidimensional User Needs

c) To cope with the newly generated knowledge, published in different forms, such as printed and electronic media on various disciplines newly generated subject areas.

d) Shrinking budgets of libraries and increasing costs of publication.

e) for better sharing of existing resources.

4. Objectives of the consortia

a) Increasing the cost benefit per subscription.

b) Promote the rational use of funds.

c) Ensure continuous subscription to the periodicals subscribed.

d) Guarantee local storage of the information acquired for continuous use by present and future users

e) Develop technical capabilities of the staff in operating and using electronic publication databases

f) Strategic alliance with institutions that have a common interest, resulting in reduced information cost, and Improved resource sharing.

In view of the above discussion, technological developments, electronic publishing of scholarly journals, the emergence of consortia and the pricing model of publisher give new opportunities for libraries to provide instant access to their collection. Moreover, while developing information collection

5. Selection Process of Consortia

- selecting a coordinating agency to deal on behalf of the entire group of participants and executing and monitoring the work.
- Identification of libraries interested in participating and agreeing to common terms and conditions.
- Identification of potential publishers to provide access under consortia purchase.
- Negotiating with publishers to get a commonly acceptable and affordable price.
- Source of funding to meet the subscription cost.
- Legal issues involved in contracts and usage of material within-the consortia.
- Informing of the usefulness / importance of the consortia to the heads of the institutions, faculty, etc., to act upon the issue.
- Identifying the necessary infrastructure for electronic access to resources.
- Issues relating to backup of databases
- Identification and selection of databases to be acquired and hosted at one place (i.e. coordinating agency).
- Documentation and training to staff.
- Access rights whether to provide direct access from publisher site or mount databases at coordinating agency.

6. Movement of Consortia based Approach

Consortium means "group of libraries". The aim of the consortia is to ensure use of information collectively from one source.

Table 1: Consortia based Commercial Services

Name of the Consortium	Web Sites
EBSCO	http://www.ebsco.com
Black well Publishers	http://www.blackwellpublishers.co.uk
Kluwer Online	http://www.kluweronline.com
NASA Technical Reports Server	http://ntrs.nasa.gov
US-Patent Full-Text and Full Image Database	http://www.uspto.gov
World Libraries on the World Wide Web	http://www.123world.com/libraries
Earthquake Information	http://www.eqnet.org

The above commercial publishers provide their service to the group of libraries.

In view of the above environment, specifically in electronic based life, in the matter of e-governance all the activities are connected with network based in LIS field. Computers are linked by telecommunication system where network offers two resources. Firstly, they offer access to people who use computer on the network by means of using internet via. e-mail, video conferencing, telnet, chatting, ftp, gopher, list serve, etc. lastly, networks permit the use of files in the form of text, graphics, sound and video, software's, subject based databases and peripherals stored on, or attached to computers in the network. The cooperative approach is proliferating in every field by helping each other and encouraging support and development by individuals and their communities.

7. Consortia Projects in Agricultural University Libraries

The general overview of the projects and programs are initiated and maintained by the agricultural university libraries all over the country and ICT based initiatives are taken for dissemination of information in libraries. The few of the projects are discussed hereunder.

7.1 Consortium for E-Resources in Agriculture (CeRA)

The Consortium for e-Resources in Agriculture (CeRA) was established in November 2007 for facilitating accessibility of scientific journals to all researchers / teachers in the National Agricultural Research System by providing access to information specially access to journals online which is crucial for having excellence in research and teaching. CeRA is promoted by Indian Council of Agricultural Research, New Delhi and funded by National Agricultural Innovation Project launched . Currently 147 institutions take benefit of CeRA.

7.1.1 Objectives of CeRA

1. To upscale the existing R & D information resource base of ICAR Institutions/Universities comparable to world's leading institutions / organizations.
2. To subscribe e-journals and create e-access culture among scientists / teachers in ICAR Institutes / Agricultural Universities.

3. To assess the impact of CeRA on the level of research publications measured through NAAS ID and Science Citation Index.

CeRA, is playing a major role while procuring print versions of journals and literature in the field of science and technology. With the advent of ICT facilities and advancement of web technology, almost all reputed international journal are available online and can easily be accessed by the researchers over the network. The ICAR is having network connectivity across the institutes and state agricultural universities, select journals could be made available over the network for the use of scientific community. Keeping this broad objectives in mind, Consortium for e-Resources in Agriculture (CeRA) at the Indian Agricultural Research Institute (IARI)

The universities library as a member of CeRA through which user can retrieve lot of free full-text foreign journals. The URL is http://www.cera.jccc.in. IP authenticated access is provided to this portal. Alphabetical as well as subject-wise list of journals is available on this portal along with search bar tool. The CeRA currently provide access to a collection of 2000 + Journals (electronic + print), from the following participating publishers are: Springer Link, Annual Reviews Inc, Elsevier Science, CSIRO, Indian Journals, Taylor and Francis, American Society Of Agronomy, Oxford University Press and Open J-Gate.

7.2 Full-text Database of Ph.D. Dissertations (Krishi Prabha):

KrishiPrabha is a full-text electronic theses repository of Indian Agricultural Doctoral Dissertations submitted by research scholars to the 45 State/Deemed Agricultural Universities during the period 2000 to 2012. This repository, listing about 8458 Doctoral Dissertations with a full text of about 6812 Dissertations has been created by Nehru Library, Ch. Charan Singh Haryana Agricultural University, Hisar with financial support from Indian Council of Agricultural Research, New Delhi under its National Agricultural Innovation Project.

7.3 AgriCat

AgriCat is the cooperative cataloging of the holding of the twelve major agricultural libraries in India including ICAR institutes and SAUs combine together. As the AgriCat is a part of e-Granth program operated and maintained by the ICAR. This collaboration of the libraries namely Indian Agricultural Research Institute (IARI), Indian Veterinary Research Institute (IVRI), University of Agricultural Sciences (UAS), G B Pant University of Agriculture and Technology (GBPUAT), Choudhari Charan Singh Haryana Agricultural University (CCSHAU), N. G. Ranga Agricultural University (ANGRAU), National Dairy Research Institute (NDRI), Central Institute of Fisheries Education (CIFE), CSKHPKV, MPKV, TANUVAS, DIPA of the ICAR institutes and SAUs combined together. The prime objective of AgriCat is to inform the user of a university who has what, which library has a material. The compilation of union catalogues includes periodical, monographs, conference proceedings, case study collections, reference materials, guide to research, microforms, audio-visual materials, dissertations and theses materials held by the participating libraries. This makes information resource sharing easily. In the age of increasing cost of publications & shrinking library budgets, consortia plays a vital role in online resource sharing & provide seamless access to electronic information.

8. Conclusion

We see all about consortia, Need of Consortia, Objectives of the consortia, Selection Process of Consortia, Movement of Consortia based Approach, Consortia Projects in Agricultural University Libraries. Finally, we would like to conclude that consortia are very important resources for providing library services to users.

References

Agnes, N., H. Sarah, O. Lisa, I. Peters, S. Anika and W. G. Stock (2013). Public libraries in the knowledge society: Core service of libraries in information world cities libri.63(4): 295-319.

Bakken, B. (2007). Young people and the Eichmann library', Scandinavian public library quarterly, 40 (1) : 6-7

Bhatt, Atual. (2012), An analytical study of the medical college libraries of Gujrat in the age of information technology, library philosophy and practice, 32 (2) : 138-145.

Chatopadhyay, T. and Ghatak, N.C. (2009), An analytical study of the dental college libraries in Kolkata in the age of information technology. SRELS journals of information management 45 (1), 45-53.

Devi, A. H. and Devi. T. P. (2005) Application of information technology with special libraries of Manipur; a case study. 3rd convention planner 2005, Silcha; Assam Univ. pp 28-37.

Divatankar, N. I. and S.S. Lokhande, (2014) the library functions and services in the ICT environment, information age, 8 (1) : 12-15.

Hicks, H. 2003, Digital libraries and education; trends and opportunities D. Lib. Magazine, 12 (2) : 78-85.

Hitchcock, S. (1997) "Citation linking improving access to online journals", Proc. 2nd ACM Int'I Conf. on digital libraries. ACM Press. New York, 1997 pp. 115-122.

Howse, David, K., Bracke, paul. J. and Keim, S.M. (2006). Technology mediator: A new role for the reference librarian? Biomedical Digital Libraries 3(10), 210-216.

http://cera.iari.res.in/index.php/en/

http://www.egranth.ac.in/AgriCat.html

Jain (2002) Jain, Vijaykumar. IT Governance Hydrabad : ICFAI university press, 2002, p. 298.

Jeevan, V. K. J and S. S. Nair (2014). Information technology adoption in libraries of Kerela : A survey of selected libraries in Thiruvananthapuram, Annual of Library and information studies, 51 (4) : 137-144.

Jena, S. and K. C.Das (2013), ICT for library professional, New Delhi, SSDN Publisher, Pp.1-10

Kaul, H. K. (2002) Knowledge centers : The key to self employment and poverty alleviation, Delnet newsletter 9 (2) : 17-18.

Krishnaphani, K. (2004), E- Government concepts and cases, Hyderabad, ICFAI, University, Press Pp. 215.

Kumar, R. P. (1994) An overview of modern technologies applications in Indian libraries, International information and library review, 26 (4) : 327-339.

Lawrence, S., C. L. Giles and K. Bollacker. (1999) digital libraries and autonomous citation indexing, IEEE computer, 32 (6) : 67-71.

Malviya, R. N. (2008). Digital library and academic society in India. Asian books, Ne Delhi 3-25.

R. D. Cameron, (1997) A universal citation database as a catalyst for reform in scholarly communication." First Monday, Arp. 1997.

Rajendra, Aparna and Joshi, Subhada M. "Use of Online Journals in University Libraries with Special Reference to Infonet: A Case Study of the University of Pune." in Library to Information Networking NACLIN 2004 edited by H. K. Kaul and S. K. Patil, November 23-26, 2004, New Delhi: Delnet, 83-89.

Ram, M. and Dr. A. K. Sharma, 2009, Library consortia in India, An over view, Library progress, 26(2) : 163-168.

Rathinasabapathy, G. (2015). Strengthening of Digital Library and Information Management under National Agricultural Research System (e-Granth): An Overview. Journal of Library, Information and Communication Technology 4.1-2 (2015): 73-78.

Rathinasabapathy, G., T. Mohana Sundari and C.Balachandran (2008). E-Journal Consortiua for Agricultural/Veterinary Universities and ICAR Institutes in India. *In:* 23rd National Seminar of IASLIC, Bose Institute, Kolkata. Spl. Pub. No.48 pp.89-96.

Singh, S.P. (2001). Computer applications in Indian institutes of technology libraries. Electronic library, 19 (2) : 92-101.

Sinha, R. P. E-Governance in India : Initiatives and issues, New Delhi concept publishing, 2006, p. 183.

Veer.D.K., Kadam Santosh & Kale, Rajesh D(2009) E-consortia : A new paradigm in online resource shairing, UGC Sponsored State Level Seminar on "Building E-Resources for Information Access in the New Millennium, At Arts College Bamkhede T.T., Tal-Shahada, Dist. Nandurbar,Pp.142-145

Vijaykumar, A. (2011) Application of information technology on libraries : an overview. International journal of digital library services, 20 (12) : 148-159.

22

Academic Libraries and E-Learning: An overview

Dr. Y.C.H.Venkateswarlu
Chief Librarian
Sri Venkateshwara College of Engineering Vidyanagara, Bangalore – 562157, Karnataka State
e-mail- ychvenkateswarlu@gmail.com

ABSTRACT

The Academic environment is also changing from formal education to distance Learning and online Learning mode because of ICT. E- Learning is more popular day by day. The academic Institute's educational System is changing and in such Case Academic Libraries have to change their practices to suit the need of e-Learning. E- Learning gives new dimensions to higher education and as well as other area of education. This paper highlights the basic concept of e-Learning, Objectives of E-Learning advantages, dis advantages, tools for e- learning and also further discuss about the participation of academic Libraries in the e-Learning. This Paper the impact of E- Learning over the academic Libraries and the challenges faced by the Libraries.

Keywords: *Academic Libraries, E-Learning, ICT*

1. Introduction

E-Learning is becoming an increasingly important part of education these days. It is the paramount importance for the professionals to provide the most viable combinations of knowledge sources that will keep their capital their employee/

worker base the most competitive possible. The scope of such learning to unpleasantly very broad, the introduction of new technologies like, e-learning into Indian scenario is ready to create many challenges, including skills, financing, capacity and many other. E-Learning is a very good technology provided by the present day technology advancement and innovations, Indian students are taking help from teachers sitting abroad in this a student from a remote area of country can complete his/her education from Capital New Delhi.

E-Learning is a new concept of virtual learning, virtual learning room, and web based education leading to establishment of virtual University with a view of extend educational opportunities for all, anywhere and at any time. Today all the information is available in electronically/digital format. Education system is faced problems such as trained and experienced teachers, lack of infrastructure and need of quality education. E-Learning could solve the problems. The future education is totally based on e-learning. E-learning (electronic learning), a wide set of applications and processes such as Web-based learning, computer-based learning, virtual classrooms and digital collaboration. It includes the delivery of content via internet, extranet (LAN/WAN), audio and videotape, satellite broadcast, interactive TV, CD-ROM and more.

E-learning makes learning interesting, interactive and fun! It has right the right blend of content (instruction) and cutting-edge technologies that offer the best benefits. It is also called as Online Education, Online Learning, Internet education, Computer-Based Training (CBT), Computer-Assisted Instruction, Virtual Education, Cyber Learning, Asynchronous Learning Networks (ALN), Web Based Training (WBT) or Learning Management Systems (LMS) etc.

2.Meaning of E-Learning:

New Zealand's Minister of Education defines e-learning as, "Learning that is enabled or supported by the use of digital tools and content. It typically involves some form of interactivity, which may include online interaction between the learner and their teacher or peers. E-learning opportunities are usually accessed via internet, through other technologies as CD-ROM are also used".E-Learning is a term that covers many approaches, which have in common the use of ICT.

E-learning is a technique of delivering educational content through Digital Interactive Television, Video-conferencing, audio-conferencing, Internet-Intranet, Worldwide Web, Video/Audio tapes, Video-on-demand, CD-ROM etc. Broadly speaking E-learning is of two types, Synchronous E-learning and Asynchronous E-learning.

- Synchronous E-Learning:

Synchronous E-Learning establishes contact between instructors and students at the real time. Example of synchronous Learning are live radio/live interactive television broadcasting videoconferencing, teleconferencing, chatting, on-line seminar etc.

- Asynchronous E-learning:

Asynchronous E-learning doesn't establish contact between instructor's students at the real time. Example includes extraction of knowledge through CD or DVD or video or audio tapes or through web pages. Correspondence through E-mail falls under this category.

3. Objectives of E-Learning:

The following are the objectives of E-learning:

- All Students and teachers will have access to information technology in their classrooms, schools, communications, and homes;
- All teachers will use technology effectively to help students achieve high academic standards;
- All students will have technology and information literacy skills.
- Research and evaluation will improve the next generation of technology applications for teaching and learning;
- Digital content and networked applications will transform teaching and learning; and
- Distance education provided the base for E-learning's development.
- 4. Advantages of E-Learning:
- E-learning provides opportunity both formal and informal learning communities;
- Learning resources can be relatively easily developed using a variety of standard packages, hence more compact and durable;
- In E-Learning, one can make use of and link into, other relevant resources available on the Internet;
- E-Learning provides flexible delivery of content material over Internet for 24x7 hours;
- Online delivery of reading materials is relatively cheap, as there are no printing and distribution costs;
- E-learning enables both one to-one and one-to-many combinations;
- Higher maintenance of content through personalized learning;
- Improved collaboration and interactivity among students. Teaching and communication techniques create an interactive online environment;

5.Disadvantages of E-Learning:

E-Learning has many disadvantages over the other methods of learning. Some of these are

- Lack of face to face conversation;
- Maintenance also very costly;
- Information and communication infrastructure is required which is capital intensive;
- Special e-learning is required to know and operate computer/Internet etc;

- E-working is power dependent;
- Technology is changing at a faster rate and its incorporation in the system is not that easy and it is costly at the same time;

6. Types of E-Learning:

- Virtual Classrooms:

  The intention of virtual classrooms is to extend the structure and service that accompany formal education programs from the campus to learners. Normal classroom are for those who may by perusing a distance education program made up entirely of online lessons. Rapid e-learning: uses tools such as Adobe captivate and Adobe Presents to reduce the time it takes to produce rich, engaging FLV learning content, while allowing more non-technical contributors.

- Online Learning:

  Learning management systems are serving as the basis for building online programs where the learning is entirely through digital mode.

- Mobile Learning:

  It takes advantage of place independent flexibility that comes from working away from the PC; it provides the opportunity to connect informal learning experiences that occur naturally throughout the day with formal learning, such as in the virtual class model using interesting programmes or online learning. Performance support systems: is simple and straight forward or much immersive, depending on need and critically of performance.

- Challenges of E-Learning:

  Determination of nature and extent of information-availability of Internet and its effective use- Use of incorporated information effectively –Access needed information effectively –Use of information ethically and legally –legal and social issues associated with the surrounding of the information.

- Corporate E-Learning:

  Corporate are using e-learning as means of communicating, training and enhancing employee value across the organization and countries, holding seminars, workshops or conferences detract employees from their work and results of such practices are at the best weak, being able to instruct employees with the job through e-learning, can prove to be extremely valuable to any business.

7. Tools of E-Learning:

There are the some following tools uses for e-learning:

- Web Blog

  A blog short for web log is a user-generated website where entries are made in journal style and displayed in a reverse chronological order. The term 'blog' is a mingling of the words web and log. It provides comments or news on a particular subject, online diaries.

- Social Bookmaking

  Social bookmaking is a web-based service to share internet bookmarks. The social bookmaking sites are a popular way to store, classify, share and search links.

- Wiki

  A wiki is a website that allows visitors to add, remove, edit and change content, without the need for registration. It also allows for linking among any number of pages. This ease of interaction and operation makes a wiki an effective tool for mass collaborative authoring.

- RSS (Really Simple Syndication)

  RSS is a web feed formats used to publish frequency updated digital content, such as blogs, news feeds and or podcasts, podcasts etc.

- Podcasting

  Podcasting is a fusion of two words i.e.iPod, Apple popular digital music player and broadcasting. Podcasting are basically digital audio programs that can be subscribed to and downloaded by listeners by RSS. It can be assesses on an array of digital audio devices like Mp3 players, desktop computer, laptops, mobile etc.

- Instant Messaging

  An Instant Messaging application allows one to communicate with another person over a network in relative privacy. There are many options like Gtalk, Sgype, Meetro, ICQ, Yahoo Messenger, MSN Messenger and AOL for instant messaging.

- Text Chat

  Internet Relay Chat (IRC) and other online chat technologies allow users to join chat rooms and communicate with many people at one, publicly. This facilities both one-to-one communication and many to-many interaction.

- Internet Forum

  Originally modeled after the real-world paradigm of electronic bulletin board of the world before internet was born, internet forums allow users to post a "topic" for others to review. Others users can view the topic and post their own comments in a linear fashion, one after the other. The above e-learning tools are a practical, inexpensive and uncomplicated method for learning online. They are available to one and all and are great propagating e-learning.

8. Academic Libraries and E-Learning

Academic libraries are considered to be the nerve centres of any academic Institution which support teaching, research and other academic programmes. Academic libraries will play an important role in development and progress of any educational system. It provides the any information to their user and fulfill their requirement related their subject, interested area, learning, teaching and research. E-learning is gave great opportunities to libraries, to used the library resources and services in support of learning, research and outreach. E-learning is the change all the traditional way of teaching methods, it is offering virtual classrooms to a without geographical boundaries and countries, it is revolutionizing change in the educational system; it is also focusing on individual rather than a group. We can say that it is totally personally attention teaching methods. E-learning is transfer to the knowledge from one to many people or groups.

Academic libraries have facilities of digital and e-learning. They can apply used their ICT's infrastructure in support of e-learning and e-research by access to electronic resources, online databases, online catalogues, e-books, e-journals, archives, digital libraries and electronic services.

The academic libraries provide these facilities to faculty, student in on/off campus. An academic library managing the services regarding e-learning, for this purpose academic libraries should establish an e-learning centre, which would support their academic curriculum with the help of faculty members and supporting staff. Academic libraries also support the e-learning with their multimedia resources, which consist of audio, video, CD-ROM, microfilm, microfiche and DVD's. Emerging communications technologies nowadays provide an opportunity to academic libraries to contain these multimedia resources. These resources will enhance access to information where anytime and anywhere 24*7 service to student, faculty members.

9. Conclusion

E-learning allows for efficient transfer of knowledge in real time process, while at the same time empowering learners with the information technology awareness and skills crucial to succeed in the present Knowledge Revolution Era. The E-learning and E-resources are dual source of nectar in information revolution in resource centers. The E-learning will be one of the hottest new technology areas within the next few years. learning educational centre and also play a vital role in e-learning education.

References:

en.wikipedia.org/wiki/e-learning.

http://www.infolibraria.com.

Kamila, K. "E-learning: The New Avenues for easy and Lifelong Learning", National Conference of AALDI on Knowledge Management in the Global Era, 21-23 April 2010: 417-422.

Promod Patel and Dr.Govind B Dave. A Study of Impact of use of IT on Academic Libraries in Uindia. International journal of Library & Information Science (IJLIS), 5(1), 2016, pp.36-40.

Rajput, P S. and at al. "E-learning in library and Information Science: Issues and Challenges in Digital Era, "National Conference of AALDI on Knowledge Management in the Global Era. 21-23 April 2010: 404-408.

Ramchandraiah, V. "ED and purpose of E-learning in Libraries: A conceptual Study," National Conference on National Reorientation of Library Services in India, 18-20 August, 2007: 183-192.

Sachdeva, G.K. and Sharma, G.S. "E-learning in Library and Information Science (LIS)", National Conference on Information Literacy Skills for College Libraries in Digital Environment (NCILSCLDE-2011), 26-27 February 2011: 169-173.

V.V.K.Suryanarayana Murthy, Analysis of User Satisfaction Levels – A Case Study on Academic Libraries. International journal of Library & Information Science (IJLIS), 5(2), 2016, pp.07-15.

23

Mapping of Global Equine Research: A Scientometric Analysis Of Publications Output DURING 1967–2016

Dr.S.Kopperundevi, Ph.D.,

Assistant Librarian
Veterinary College & Research Institute
Tamil Nadu Veterinary and Animal Sciences University
Tirunelveli – 627 385, Tamil Nadu
e-mail: devielangowins@gmail.com

ABSTRACT

Objective: *This study analyses the research output on Equine during the period from 1967 till 2016 i.e. 50 years. It analyses the growth, rank and global publications share, contribution of major countries. It also analyses the characteristics of most productive authors.* ***Materials and Methods:*** *CAB Direct Online database has been used to retrieve the data on publication output on equine research.* ***Results:*** *The present study revealed that 67,195 papers were published during the period 1967-2016 and the growth of publications. It has been observed that during the year 1977-1986, there was an increase of 231.46% in equine research publications and a steady growth is* *recorded till 2016. The publication share of the top 10 countries varies from 0.98 to 4.89% in global publication output during the past five decades (1967 – 2016). United States scores the first position with a publication share of 4.89% and Germany comes second with a publication share of 1.99%. India ranks 9th position among these top 10 countries in equine research by contributing 1.01% of the publications share in world equine research during 1967-2016. Top 20 authors, who had published 150 and more papers in equine research during 1967-2016, have been identified as the most productive. The combined contribution of these 20 authors accounts for 5.85% share (3,933 papers) in the global output in equine research, with average productivity of 196.65 papers per author.* ***Conclusion:*** *It is concluded that the equine research output was steadily increased till 2012 and it is declining since 2013 as the number of publications is decreasing. Indian contribution on equine research needs to be improved. So, the stakeholders of Indian equine industry and policy makers may support for more research on equines.*

Keywords: *Equine, Horse, Veterinary medicine, Animal Science, Livestock, Large Animal, CAB Direct, Scientometrics, Publication Output*

1. Introduction

Horses are present throughout the world and have been with humans throughout history. It is surely the 'aristocrat' of animals domesticated by man. Horse is an important animal in India also. It belongs to Equus genus of mammals in the family Equidae which also includes horses, asses, and zebras. India possesses 1.17 million equines. Major population of equidae comprising donkeys, mules and ponies that provide livelihood to the rural societies living in arid, semi-arid and hilly regions. While equines are used for transport and draught in Himalayas, a small population of equines is used in army, police, border security force, racing industry and sports.

In recent times, because of their power, agility, gracefulness and speed, horses are mostly used for personal pleasure and in sport competitions. In recent years the globalization of horses has been widely recognized being developed as sports animal. The trade, breeding and sports significantly attracted the attention of people. Like other species of animals, horses are also an important component of global biodiversity. The equine industry currently has a great economic effect in countries like United States, Australia and New Zealand as it generates huge amounts of revenue.

2. Equine Research

Since equine industry has its impact on the global economy, much importance is given to equine research across the globe. Even in India, the need for having a special research institute for equines was felt by the Indian Council of Agricultural Research (ICAR) and it has established the 'National Research Centre on Equines" (NRCE).

The common diseases affecting equines health are equine infectious anemia, equine viral arteritis, equine herpes virus and contagious equine metritis. So, research into equine medicine has developed a variety of vaccines and antitoxins to treat horse infectious diseases including botulism, rabies, diphtheria and infection from H5N1 influenza. In India, almost all veterinary colleges has the facility to treat any type of health problems of equines including surgery.

3. Objectives

The main objective of this study was to analyze the research performance on equine in the global context, as reflected in its publication output during 1967-2016 in CAB Direct Online Database. In particular, the study focuses on analysing the following

i equine research output, its growth, rank and global publications share

ii the contributions by sub-fields of equine science

iii the publication productivity of leading authors

iv the publication productivity of leading nations

v preferred medium of communication

vi leading journals publishing equine research; and

vii preferred language of communication

4. Review of Literature

The review of literature revealed that not much work has been done on scientometric analysis on equine research and few scientometric studies carried out on other species are reviewed here. A scientometric study was undertaken by Rathinasabapathy and Rajendran (2010) and compared the R&D output on 'Buffaloes' by the researchers in India and Pakistan. The chosen study period is 55 years (1955-2009) and CAB Direct Online was the source database for this research. It was reported that a total of 9,096 and 706 publications were published by the scientists of India and Pakistan respectively on buffalo research and India is the top producing country with 9,096 papers (92.80%) followed by Pakistan with 706 papers (7.20%).

Malathi and Ravi (2012) have studied the growth and collaboration trends in livestock research in India as reflected in Science Citation Index during 1999-2010 and found that a total of 600 papers in SCI covered journals. The scientometric study undertaken by Rathinasabapathy (2012) on goat research based on CAB Direct Online database revealed that there were 31,413 publications on goat and 84.75%. Rathinasabapathy and Kopperundevi (2013) have undertaken a scientometric study and mapped bovine mastitis research covering the publications during the period 1962 –2013, which was a pioneer work on scientometric analysis of animal disease.

Kinth et al (2013) have conducted a scientometric study and mapped global zebrafish research.A scientometric assessment on camel research by Rathinasabapathy and Rajendran (2013) revealed that a total of 4923 publications were indexed by CAB Direct during the period 1963-2012 and India was the top producing country with 354 papers which was 7.19%. T.K.Gahlot, an Indian was the prolific author as he has published 173 papers (3.51% of global publication output). Similar study on global broiler research output was carried out by Rathinasabapathy (2013) which reported a detailed study on the broiler research carried out across the globe. The quantitative growth and development of world literature on 'duck research' was traced by G.Rathinasabapathy (2013) and reported that there were 6,836 publications during the period 1970-2012and China was the top producing country with 539 papers.

Sivakami (2015) has carried out a Scientometric analysis of Swine Infulenza Research Output from 1990 to 2013. A scientometric study to map the dynamics of 100years of global sheep research was carried out by Rathinasabapathy et al (2014). Alfred and Kipanyula (2016) have conducted a scientometric mapping of veterinary research at Sokoine University of Agriculture, Tanzania.

5. Materials and Methods

5.1 Source of Data

This study was based on the publication data on equine, retrieved from the CAB Direct Online database for 50 years (1967 – 2016). The search strategy/keyword used to retrieve the data on equine was "equine" and "horse" in Article field using the Advanced Search facility. To retrieve the publications on equine, the following string was used in CAB Direct Online database:

(title:(equine) OR title: (horse)) AND Published between (1967) AND (2016)

6. Results

6.1 Growth of World Equine Research Output

The present study revealed that 67,195 papers were published during the period 1967-2016 and the growth of publications is furnished in Table-1. It has been observed that during the year 1977-1986, there was an increase of 231.46% in equine research publications and a steady growth is recorded till 2016.

Table-1 Growth of World Equine Research Output

S. No.	Block Year	Total Publications	Growth %
1	1967 – 1976	4,920	-
2	1977 - 1986	11,388	231.46%
3	1987 – 1996	14,652	28.66%
4	1997 - 2006	15,584	6.36%
5	2007 - 2016	20,651	32.51%
	Total	67,195	-

While during the past five decades, there was steady growth in equine research publications, a close watch on the growth of publications revealed that during the last four years i.e from 2013 to 2016, there is a declining trend on equine research as the number of publications is slowly decreasing as the total number of publications during 2013 was 2,305 and it came down to 2,186 during 2014 and the year 2015 saw again a decline with 1,974 publications and during 2016, it came down further to 1,899. This declining trend needs to be taken care of by the stakeholders of equine research.

6.2 Contribution of various sub-fields in equine research

Analysis of equine research output in the context of different subjects revealed that 21.87% (14,693 papers) of the total research output have been in horse diseases, followed by diagnosis (14.22% share and 9,557 papers), infections (9.84% share and 6,617 papers), surgery (7.86% share and 5,283 papers), bacterium (7.48% share and 5,032 papers), parasites (7.09% share and 4,766 papers), race horses (6.44% share and 4,328 papers), viral diseases (6.09% share and 4,095 papers), diagnostic tests (5.64% share and 3,794 papers) mares (5.57% share and 3,744 papers). The top ten sub-fields in equine research are depicted in Figure-1.

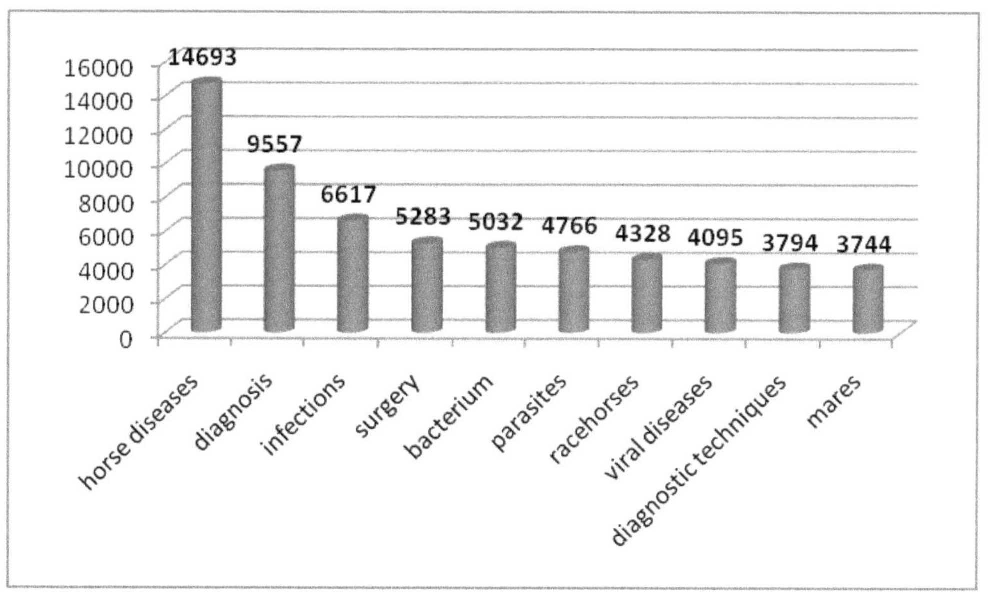

Fig.1 Contribution of various sub-fields in equine research

6.3 Publication Share and Ranking of Countries

The publication share of the top 10 countries varies from 0.98 to 4.89% in global publication output during the past five decades (1967 – 2016). United States scores the first position with a publication share of 4.89% and Germany comes second with a publication share of 1.99%. Brazil, UK, France, Poland, Australia and Nordic Countries are placed from third to eighth positions with publications share ranging from 1.86% to 0.98%. India ranks 9th position among these top 10 countries in equine research by contributing 1.01% of the publications share in world equine research during 1967-2016. The details are furnished in Table-2.

Table-2 Ranking of Countries

S. No.	Country	Total Publications	% of Share
1	USA	3,285	4.89
2	Germany	1,338	1.99
3	Brazil	1,250	1.86

4	UK	1,221	1.82
5	France	905	1.35
6	Poland	837	1.25
7	Australia	828	1.23
8	Nordic Countries	778	1.16
9	India	680	1.01
10	Italy	656	0.97
11	Africa South of Sahara	576	0.86
12	Japan	563	0.83
13	Canada	497	0.82
14	USSR	469	0.69
15	Sweden	357	0.53
16	China	333	0.49
17	Spain	320	0.48
18	Netherlands The	296	0.44
19	Argentina	286	0.43
20	Romania	280	0.42
	Total	15,755	23.44

6.4 Research profile of prolific authors engaged in equine research

Top 20 authors, who had published 150 and more papers in equine research during 1967-2016, have been identified as the most productive and their research performance is presented in Table-3. The combined contribution of these 20 authors accounts for 5.85% share (3,933 papers) in the global output in equine research, with average productivity of 196.65 papers per author. Eight authors have contributed higher publications than the group's average. The details are furnished in Table-3

Table-3: Research Profile of Prolific Authors

S. No.	Author	Total Publication	% of contribution
1	Wade J.F.	380	0.57
2	Mcllwraith, C.W.	262	0.39
3	Deegen, E	233	0.35
4	Denoix, J.M.	231	0.35
5	Mumford, J.A.	214	0.33
6	Divers, T.J.	203	0.30
7	Tobin, T	202	0.30
8	Rossdale, P.D.	197	0.29
9	Schumacher, J	193	0.29
10	Moore, J.N.	191	0.29

11	Houghton, E.	180	0.28
12	Hinchcliff, K.W.	168	0.25
13	Muir, W.W.	167	0.24
14	Allen, W.R.	164	0.24
15	Dixon, P.M.	164	0.24
16	Van Weeren, P.R.	164	0.24
17	Lindner, A.	159	0.23
18	Geor, R.J.	158	0.23
19	Plowright, W.	153	0.22
20	Timoney, P.J.	150	0.22
	Total	3,933	5.85

6.5 Patterns of research communication

The equine research output published in 20 productive journals is presented in Table-3. All these 20 journals are published by foreign countries and none of the Indian journal is in the top 20 journal list. The cumulative publications output of these 20 most productive journals contributes 29.66% share (19,931 papers) in the global output during 1967-2016.

Table-4 Productive Journals – Journals receiving largest number of papers

S. No.	Journal Title	No. of Papers	% of contribution
1	Equine Veterinary Journal	3,310	4.93
2	American Journal of Veterinary Research	2,221	3.30
3	Journal of the American Veterinary Medical Association (JAVMA)	1,792	2.67
4	Journal of Equine Veterinary Science	1,623	2.42
5	Veterinary Record	1,402	2.08
6	Pferdeheilkunde	1,279	1.90
7	Veterinary Surgery	1,153	1.73
8	Equine Veterinary Education	986	1.48
9	Praktische Tierarzt	658	0.98
10	Veterinary Clinics of North America: Equine Practice	617	0.91
11	Australian Veterinary Journal	605	0.90
12	Equine Practice	601	0.89
13	Veterinary Journal	564	0.84
14	Journal of Veterinary Internal Medicine	533	0.79
15	Compendium on Continuing Education for the Practicing Veterinarian	446	0.66
16	Research in Veterinary Science	443	0.67
17	Canadian Veterinary Journal	427	0.63

18	Theriogenology	426	0.63
19	Journal of Veterinary Pharmacology & Therapeutics	425	0.63
20	Veterinary Immunology and Immunopathology	420	0.62
	Total	19,931	29.66

6.6 Preferred form of research communication

The equine research output publications by the authors revealed that 88.80% are journal articles followed by conference papers 10.43% and the remaining publications in other formats *viz.*, book chapters, thesis, editorial, abstract, etc. The details are depicted in Fig.1

7. Preferred Languages

The study found that English is the preferred language by the authors of equine research to publish their research papers as 72.74% (48,881 papers) were published in English followed by German 7.70% (5,180 papers), French 4.32% (2,905 papers), Portuguese 2.86% (1,923 papers) and Spanish 1.99% (1,341 papers). The top 10 languages preferred by the authors are depicted in Figure-2.

8. Findings of the Study

- The present study revealed that 67,195 papers were published during the period 1967-2016 and the growth of publications.
- It has been observed that during the year 1977-1986, there was an increase of 231.46% in equine research publications and a steady growth is recorded till 2016.
- There is a declining trend on equine research during the last four years as the number of publications is slowly decreasing as the total number of publications during 2013 was 2,305 and it came down to 2,186 during 2014 and the year 2015 saw again a decline with 1,974 publications and during 2016, it came down further to 1,899. This declining trend needs to be taken care of by the stakeholders of equine research.
- Analysis of equine research output in the context of different subjects revealed that 21.87% (14,693 papers) of the total research output have been in horse diseases, followed by diagnosis (14.22% share and 9,557 papers), infections (9.84% share and 6,617 papers), surgery (7.86% share and 5,283 papers) and, bacterium (7.48% share and 5,032 papers).
  - The publication share of the top 10 countries varies from 0.98 to 4.89% in global publication output during the past five decades (1967 – 2016).
  - United States scores the first position with a publication share of 4.89% and Germany comes second with a publication share of 1.99%. Brazil, UK, France, Poland, Australia and Nordic Countries are placed from third to eighth positions with publications share ranging from 1.86% to 0.98%.

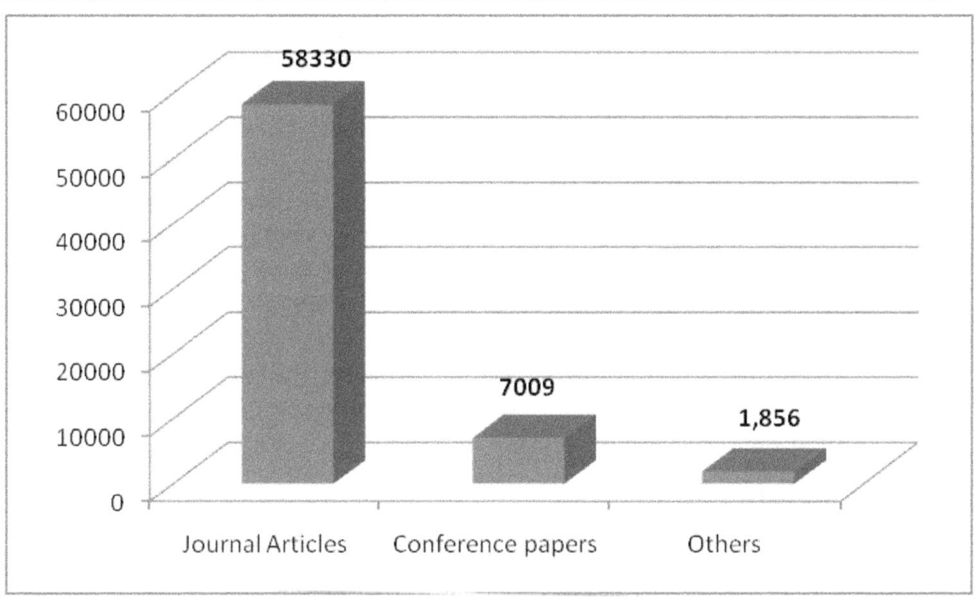

Fig.1 Preferred form of research communication

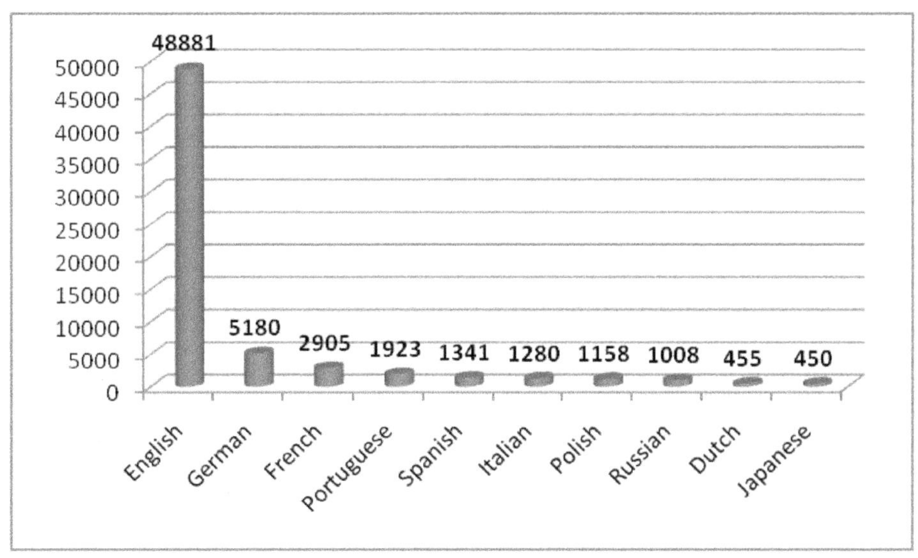

Fig.2 Preferred Languages

- India ranks 9th position among these top 10 countries in equine research by contributing 1.01% of the publications share in world equine research during 1967-2016.
- Top 20 authors, who had published 150 and more papers in equine research during 1967-2016, have been identified as the most productive.

- The combined contribution of these 20 authors accounts for 5.85% share (3,933 papers) in the global output in equine research, with average productivity of 196.65 papers per author. Eight authors have contributed higher publications than the group's average. The details are furnished in Table-3
- The equine research output published in 20 productive journals is published by foreign countries and none of the Indian journal is in the top 20 journal list. The cumulative publications output of these 20 most productive journals contributes 29.66% share (19,931 papers) in the global output during 1967-2016.
- The equine research output publications by the authors revealed that 88.80% are journal articles followed by conference papers 10.43% and the remaining publications in other formats.

8. Conclusion

It is concluded that the equine research output was steadily increased till 2012 and it is declining since 2013 as the number of publications is decreasing. This declining trend needs to be taken care of by the stakeholders of equine research. India is in 9th position in terms of number of research publications on equine research with 610 publications (1.01%) and none of the Indian journals are in the top 20 most productive journals publishing equine research. Though there is an ICAR Institute exclusively for Equine research, the number of papers published from India is very limited. Further, a journal named "Centaur" was already published which published only equine research publications and it has been stopped by the publishers. So, the stakeholders of Indian equine industry and policy makers may support publication of an exclusive journal for equine science.

References

Alfred S., and Maulilio J. Kipanyula. (2016). Scientometric mapping of veterinary research at Sokoine University of Agriculture, Tanzania. International Journal of Digital Library Services Vol. 6 (3) available at http://www.ijodls.in/uploads/3/6/0/3/3603729/19-30.pdf.

Gupta, B. M., et al. (2015). World camel research: a scientometric assessment, 2003–2012. Scientometrics 102.1 (2015): 957-975.

Kinth, Priyamvadah, Gopalakrishnan Mahesh, and Yatish Panwar (2013). Mapping of zebrafish research: a global outlook. Zebrafish 10.4 (2013): 510-517.

Malathi, P., and S. Ravi. "Growth and Collaboration Trends in Livestock Research in India: A Scientometrics Analysis." Indian Journal of Information Sources & Services (IJISS) 2.1 (2012).

Rathinasabapathy, G. (2010). Scientometric dimensions of buffalo research in SAARC countries. In *Proceedings of the VI International Conference of Webometrics, Informatics & Scientometrics, COLLNET* (pp. 90-97).

Rathinasabapathy, G. (2012). Global broiler research output: A Scientometric Study based on CAB Direct Online. Journal of Library, Information and Communication Technology 5.3-4 (2015): 107-112

Rathinasabapathy, G. (2012). Scientometric analysis of publications on goat research: A profile based on CAB Direct Online. Indian Journal of Agricultural Library and Information Services, Vol. 28 (2), pp.35-39

Rathinasabapathy, G. (2013). A Scientometric Study on Duck Research as reflected in CAB Direct Online." Journal of Library, Information and Communication Technology 5.1-2: 61-67

Rathinasabapathy, G. (2013). Bluetongue Research: A scientometric Profile Based on CAB Direct Online. Asian Journal of Library and Information Science, 5 (1-2): 47-54.

Rathinasabapathy, G. and L. Rajendran (2013). Mapping of world-wide camel research publications: A scientometric analysis. Journal of Library, Information and Communication Technology (2013).

Rathinasabapathy, G. and Rajendran, L. (2010). A scientometric study on buffalo research in India and Pakistan: A profile based on CAB Direct Online. *Asian Journal of Library and Information Science*. Available at http://arizona. openrepository. com/ arizona/ handle/10150/299577

Rathinasabapathy, G., and S. Kopperundevi.(2014). Mapping of Bovine Mastitis Research: A Scientometric Analysis of Research Output(1962-2013). Journal of Library, Information and Vol. 6 : Issue (1-2) Communication Technology (JLICT) June, 2014, pp. 61-69.

Rathinasabapathy, G., S. Kopperundevi, and L. Rajendran (2014). Mapping the Dynamics of 100 Years of Global Sheep Research: A Scientometric Analysis (1914-2013). Asian Journal of Library and Information Science 6.1-2: 39-50

Rupak, K. (2010). Global Horse Population with respect to Breeds and Risk Status. Available at http://stud.epsilon.slu.se/7676/17/khadka_r_150305.pdf

Sivakami, N. "A Scientometric analysis of Swine Infulenza Research Output from 1990 to 2013." (2015). Available at http://ir.inflibnet.ac.in:8080/ jspui/ bitstream/ 10603/ 106502/1/1%20-%20front%20pages-1.pdf

24

The Semantic Web and Linked Data: An overview

Sangeeta S. Bawaskar

Research Student, Department of LIS,
Dr. Babasaheb Ambedkar Marathwada University, Aurangabad.

Dr. Vaishali Khaparde

Professor and Head, Department of LIS,
Dr. Babasaheb Ambedkar Marathwada University, Aurangabad

ABSTRACT

The term Linked Data refers to a set of best practices for publishing and connectingstructured data on the Web. These best practices have been adopted by an increasingnumber of data providers over the last three years, leading to the creation of a global dataspace containing billions of assertions - the Web of Data. In this article we present theconcept and technical principles of Linked Data, and situate these within the broader contextof related technological developments. We describe progress to date in publishing LinkedData on the Web, review applications that have been developed to exploit the Web of Data,and map out a research agenda for the Linked Data community as it moves forward. linked Data and the Semantic Web,Definition of Semantic Web and Link Data, Motivation, Existing Work, Linked Datasets as a Proxy for the Semantic Web, Gauging the Semantic Web, What is Linked Data?, Principles, Linked open data, The Linked Data Technology Stack , Link Generation,

Metadata, Linked Data Browsers, Conclusion.

Keywords: *Linked Data, Web of Data, Semantic Web, Data Sharing, Data Exploration*

1. Introduction

The Semantic Web is related to the Semantic Web Linked Data, ontology engineering, and so forth. A unique feature of the SW is that it adopts an open and transparent review process , in which the reviews from reviewers, the authors' response letters, multiple versions of the revised manuscripts, as well as the editor's de- cisions are publicly available on the web page. This process creates a rich dataset that can be used in a variety of studies. For example, it can be used to learn about researchers, publications, trending topics, and popular paper categories. This paper provides a formal documentation of this SW dataset. Particularly, we describe the novelty of this dataset, i.e., how it distinguishes itself from other general bibliographic datasets, and why it is of interest to the Se-mantic Web community. The availability of this dataset, including its SPARQL endpoint, bulk download URL, as well as a scientometrics portal have been pro- vided. We also discuss some design considerations of the dataset in this paper.

2. Literature review

2.1 Semantic Web and Link Data

Most people thought WWW wouldn't become successful as there a lot of problems like, who will upload the data, who will manage it, who will fix the issues etc. But the widespread adoption of WWW has been on a planetary level and just about everything has a webpage for it. The Web infrastructure currently is a distributed network of interlinked webpages with UniqueResource Locators. This helps to categorize webpages of a particular niche and identify them.The idea of Semantic Web is to push the very same infrastructure, where the linking of resourcesis on the data level. Semantic Web is based on the idea of Smart Data

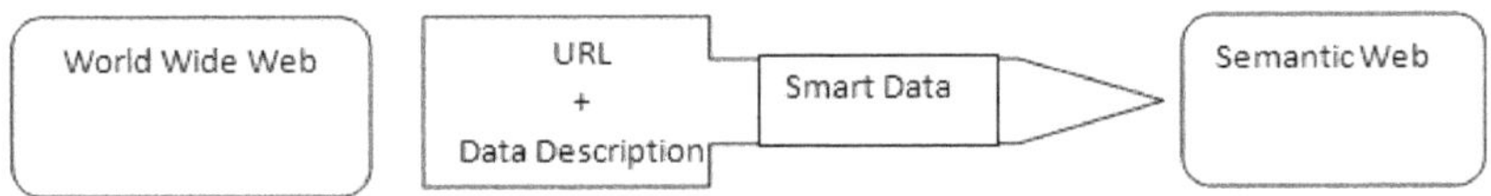

Figure 1: From Web to Semantic Web

Smart Data is interlinked data that allows not only humans to use the information, but machines too. Even if each entity of the data is held by individual organization, since they are al interlinked, it could make more meaning [10]. Sir Tim Berners-Lee believed that when interlinked data could also have the property of self-description, it would lead to Semantic Web.

Linked Linguistic Open Data (LLOD) is heavily dependent on metadata, and anyconsideration thereof would require an examination of its standards. A brief history of the topicof linguistic annotation can be found in Palmer &Xue (2013). Bird & Simons (2003a) and Ide,Romary, & de la Clergerie (2004) proposed sets of best practices for linguistic annotations, whileSimons, Bird, &Spanne (2008) offered a more recent set of recommendations that specificallysuggested language codes from ISO 639-31 be used in metadata. Ide &Pustejovsky (2010)suggested a list of best practices for language technology metadata, focusing heavily on the workof the OLAC and European Languages Resource Association (ELRA). Gracia, Montiel-Ponsoda, Cimiano, Gómez-Pérez. Buitelaar, & McCrae (2012) considered the issue of Linked Data being stored in different languages, and suggested that techniques such as ontology localization, ontology mapping, and cross-lingual ontology-based information access andpresentation would help prevent information from being locked up in linguistic data silos. Gayo,Kontokostas, & Auer (2013) presented a set of best practices for multilingual linked open data,and point out that SPARQL queries can be improved if tags are identified by language.

3. What is linked Data and the Semantic Web?

The collection of Semantic Web technologies (RDF, OWL, SKOS, SPARQL, etc.) provides an environment where application can query that data, draw inferences using vocabularies, etc ... This collection of interrelated datasets on the Web can also be referred to as Linked Data

3.1 Definition of Semantic Web and Link Data

By Vangie Beal in Semantic Web terminology, Linked Data is the term used to describe a method of exposing and connecting data on the Web from different sources. Currently, the Web uses hypertext links that allow people to move from one document to another.

3.2 Motivation

Developments in the last twelve months demonstrate that the Semantic Webhas arrived. Initiatives such as the Linking Open Data community project1 arepopulating the Web with vast amounts of distributed yet interlinked RDF data.Anyone seeking to implement applications based on this data needs basic informationabout the system with which they are working. We will argue thatregarding the size of the Semantic Web, there is more to find than the sheernumbers of triples currently available; we aim at answering what seems to be arather a simple question: What is the size of the Semantic Web?

3.3 Existing Work

On the Web of Documents, typically the number of users, pages or links are usedto gauge its size [Broder et al. 00, Gulli and Signorini 05]. However, Web links(@href) are untyped, hence leaving its interpretation to the end-user [Ayers 07].On the Semantic Web we basically deal with a directed labelled graph where afair amount of knowledge is captured by the links between its nodes.

3.4 Linked Datasets as a Proxy forthe Semantic Web

The reference test data set (RTDS) we aim to use should be able to serve asa good proxy for the Semantic Web, hence it (i) must cover a range of differenttopics (such as people-related data, geo-spatial information, etc.), (ii)must be strongly interlinked, and (iii) must contain a sufficient number of RDFtriples (we assume some millions of triples sufficient). As none of the availablealternatives—such as the Lehigh University Benchmark dataset2, SemanticWikis (such as [V¨olkel et al. 06]) or embedded metadata—exhibit the desiredcharacteristics, the Linking Open Data datasets were chosen as the RTDS.

4. Gauging the Semantic Web

In order to find metrics for the Semantic Web we examine its properties by inducingfrom the LOD dataset analysis. One possible dimension to asses the sizeof a system like the Semantic Web is the data dimension. Regarding data on theSemantic Web, we roughly differentiate into: (i) the schema level (cf. ontologydirectories, such as OntoSelect5), (ii) the instance level, i.e. a concrete occurrenceof an item regarding a certain schema (see also [Hausenblas et al. 07]), andthe actual interlinking: the connection between items; represented in URIrefsand interpretable via HTTP. This aspect of the data dimension will be the maintopic of our investigations, below.

5. What is Linked Data?

In summary, Linked Data is simply about using the Web to create typed links between datafrom different sources. These may be as diverse as databases maintained by two organisations in different geographical locations, or simply heterogeneous systems within one organisation that, historically, have not easily interoperated at the data level.Technically, Linked Data refers to data published on the Web in such a way that it ismachine-readable, its meaning is explicitlydefined, it is linked to other external data sets,and can in turn be linked to from external data sets.While the primary units of the hypertext Web are HTML (HyperText Markup Language) documents connected by untyped hyperlinks, Linked Data relies on documents containingdata in RDF (Resource Description Framework) format (Klyne and Carroll, 2004). However,rather than simply connecting these documents, Linked Data uses RDF to make typedstatements that link arbitrary things in the world. The result, which we will refer to as theWeb of Data, may more accurately be described as *a web of things in the world, describedby data on the Web.*

Berners-Lee (2006) outlined a set of 'rules' for publishing data on the Web in a way that all published data becomes part of a single global data space:

1. Use URIs as names for things
2. Use HTTP URIs so that people can look up those names
3. When someone looks up a URI, provide useful information, using the standards (RDF, SPARQL)
4. Include links to other URIs, so that they can discover more things

These have become known as the 'Linked Data principles', and provide a basic recipe for publishing and connecting data using the infrastructure of the Web while adhering to its architecture and standards.

Principles

Tim Berners-Lee outlined four principles of linked data in his "Linked Data" note of 2006,[2] paraphrased along the following lines:

1. Use URIs to name (identify) things.
2. Use HTTP URIs so that these things can be looked up (interpreted, «dereferenced").
3. Provide useful information about what a name identifies when it's looked up, using open standards such as RDF, SPARQL, etc.
4. Refer to other things using their HTTP URI-based names when publishing data on the Web.

Tim Berners-Lee gave a presentation on linked data at the TED 2009 conference.[3] In it, he restated the linked data principles as three «extremely simple» rules:

1. All kinds of conceptual things, they have names now that start with HTTP.
2. If I take one of these HTTP names and I look it up...I will get back some data in a standard format which is kind of useful data that somebody might like to know about that thing, about that event.
3. When I get back that information it's not just got somebody's height and weight and when they were born, it's got relationships. And when it has relationships, whenever it expresses a relationship then the other thing that it's related to is given one of those names that starts with HTTP.

5.2 Linked open data

Linked open data is linked data that is open content. Tim Berners-Lee gives the clearest definition of linked open data in differentiation with linked data.

Linked Open Data (LOD) is Linked Data which is released under an open licence, which does not impede its reuse for free. —*Tim Berners-Lee, Linked Data*

5.2 The Linked Data Technology Stack

Linked Data relies on two technologies that are fundamental to the Web: Uniform ResourceIdentifiers (URIs) (Berners-Lee et al., 2005) and the HyperText Transfer Protocol (HTTP)(Fielding et al., 1999). While Uniform Resource Locators (URLs) have become familiar asaddresses for documents and other entities that can be located on the Web, UniformResource Identifiers provide a more generic means to identify any entity that exists in theworld.Where entities are identified by URIs that use the *http://* scheme, these entities can belooked up simply by dereferencing the URI over the HTTP protocol. In this way, the HTTPprotocol provides a simple yet universal mechanism for retrieving resources that can be serialised as a stream of bytes (such as a photograph of a dog), or retrieving descriptions ofentities that cannot themselves be sent across the network in this way (such as the dogitself).

5.3 Link Generation

RDF links allow client applications to navigate between data sources and to discover additional data. In order to be part of the Web of Data, data sources should set RDF links torelated entities in other data sources. As data sources often provide information about largenumbers of entities, it is common practice to use automated or semi-automated approachesto generate RDF links.

5.4 Metadata

Linked Data should be published alongside several types of metadata, in order to increaseits utility for data consumers. In order to enable clients to assess the quality of publisheddata and to determine whether they want to trust data, data should be accompanied withmeta-information about its creator, its creation date as well as the creation method (Hartig,2009). Basic provenance meta-information can be provided using Dublin Core terms or theSemantic Web Publishing vocabulary (Carroll et al., 2005). The Open Provenance Model(Moreau et al., 2008) provides terms for describing data transformation workflows. In (Zhaoet al., 2008), the authors propose a method for providing evidence for RDF links and fortracing how the RDF links change over time

5.5 Linked Data Browsers

Tabulator Browser (MIT, USA)

- Marbles (FU Berlin, DE)< Disco Hyperdata Browser (FU Berlin, DE)
- Open Link RDF Browser (OpenLink, UK)
- Zitgist RDF Browser (Zitgist, USA)
- Humboldt (HP Labs, UK)
- Fenfire (DERI, Irland)

6. Conclusions

Linked Data principles and practices have been adopted by an increasing number of dataproviders, resulting in the creation of a global data space on the Web containing billions ofRDF triples. Just as the Web has brought about a revolution in the publication andconsumption of documents, Linked Data has the potential to enable a revolution in how data is accessed and utilised. The success of Web APIs has shown the power of applications thatcan be created by mashing up content from different Web data sources. However, mashup developers face the challenge of scaling their development approach beyond fixed,predefined data silos, to encompass large numbers of data sets with heterogeneous datamodels and access methods. In contrast, Linked Data realizes the vision of evolving theWeb into a global data commons, allowing applications to operate on top of an unboundedset of data sources, via standardised access mechanisms. If the research challengeshighlighted above can be adequately addressed, we expect that Linked Data will enable asignificant evolutionary step in leading the Web to its full potential.

Reference

Ayers D. Evolving the Link. IEEE Internet Computing, 11(3):94–96, 2007.

Bizer C., Heath T., Ayers D., and Raimond Y. Interlinking OpenData on the Web (Poster). In 4th European Semantic Web Conference(ESWC2007), pages 802–815, 2007.

Broder A., Kumar R., Maghoul F., Raghavan P., Rajagopalan S.,Stata R., Tomkins A., and Wiener J. Graph structure in the Web. Computer Networks:The International Journal of Computer and Telecommunications Networking,33(1-6):309–320, 2000.

Ding L. and Finin T. Characterizing the Semantic Web on theWeb. In 5th International Semantic Web Conference, ISWC 2006, pages 242–257,2006.

Ding L., Zhou L., Finin T., and Joshi A. How the Semantic Web isBeing Used:An Analysis of FOAF Documents. In 38th International Conference onSystem Sciences, 2005.

Esmaili K. and Abolhassani H. A CategorizationScheme for Semantic Web Search Engines. In 4th ACS/IEEE International Conferenceon Computer Systems and Applications (AICCSA-06), Sharjah, UAE, 2006.

Finin T., Ding L., Pan R., Joshi A., Kolari P., Java A., and Peng Y.Swoogle: Searching for knowledge on the Semantic Web. In AAAI 05 (intelligentsystems demo), 2005.

Gulli A. and Signorini A. The Indexable Web is More than11.5 Billion Pages. In WWW '05: Special interest tracks and posters of the 14thinternational conference on World Wide Web, pages 902–903, 2005.

Hausenblas M., Slany W., and Ayers D. A Performance andScalability Metric for Virtual RDF Graphs. In 3rd Workshop on Scripting for theSemantic Web (SFSW07), Innsbruck, Austria, 2007.

Tummarello G., Delbru R., and Oren E. Sindice.com: Weavingthe Open Linked Data. In The Semantic Web, 6th International Semantic Web Conference,2nd Asian Semantic Web Conference, ISWC 2007 + ASWC 2007, pages552–565, 2007.

V¨olkel M., Kr¨otzsch M., Vrandecic D., Haller H., and Studer R. SemanticWikipedia. In 15th International Conference on World Wide Web, WWW2006, pages 585–594, 2006.

Wang T. D. Gauging Ontologies and Schemas by Numbers. In 4th InternationalWorkshop on Evaluation of Ontologies for the Web (EON2006), 2006.

25

e-Resources in Andhra Pradesh Junior College Libraries: An Overview

Dr. T. Sreenivasa Rao[1] and Dr. D. Bhaskar[2]

[1]Assistant Professor in LIS, ANGR Agricultural University Library Lam, Guntur, Andra Pradesh. e-mail: sreeni.apj@gmail.com

[2]Librarian, ARJC Junior College, Venkatagiri, Nellore, Andra Pradesh

ABSTRACT

To encourage students' use of the library, and in particular of its electronic resources, we need to understand what factors encourage students to seek out information in the library setting. This paper looks at the role self-efficacy plays in their search for information and use of the library's electronic resources and computer use were analyzed and correlated with their self-efficacy. The use of the library correlated to the students' use of the library's electronic resources, express an interest in learning about the library's electronic resources will be more likely to have higher self-efficacy.

Key Words: *e-resources, self-efficacy, digital environment, transformation, Andra Pradesh*

1. Introduction

Information professionals have long sought to comprehend what factors are relevant in encouraging a person to seek out information. More recently, a particular focus of inquiry has been on those factors that play a role in deciding to use the library and its resources as a place to seek information (whether physically or virtually) as opposed to just surfing the Internet. These inquiries assume an even greater importance in light of the fact that more people are using the Internet to find information they need, information that is unmediated by the library.

Library is a social institution and facing a variety of new challenges from multiple sectors of the information society due to fast changing, expanding diverse, global digital information environment. The major challenges are:

- Information explosion
- ICT revolution
- Explosive growth and uses of Internet and web sources
- Dwindling library budget
- Escalating cost of printed documents
- Heightened level of users expectation
- Changing education and learning environment
- Evolution of virtual educational institution

These challenges have called for reorientation, reengineering and transformation of traditional libraries into e-libraries having e-resources to provide the facility of round the clock remote access.

2. What is *e*-Resource?

The resources which are available in electronic format are known as e-resources. According to AACR2 (2005) electronic resources mean "the materials (data and/ or programmes) encoded for manipulation by a computerized device. The materials may require the use of peripherals directly connected to a computerized device or a connection to a computer network (e.g. the Internet)". International Standard Bibliographic Description for Electronic resources (ISBD-ER) defines e-resources as all those materials codified for computer elaboration including material which requires the use of peripheral. Thus e-resources are accessible/ readable through an electronic device. These are available online via internet and offline via storage devices.

3. Types of Electronic Resources

The library resources are found in various formats such as e-books, e-journals, online databases, CD-ROMs, reference sources etc. The e-resources are classified as formal and informal.

4. The Formal E-Resources

- **Indexing and abstracting databases:** A large number of indexing and abstracting information services of the world have created their own databases, which can be searched from any part of the world on payment basis to retrieve necessary information. There are a number of vendors that possess plenty of on line databases.

- **Full-text databases**: These resources provide full-text of the document apart from its bibliographical information. Now a day's various publishers are providing access to full-text/databases through the Internet like American chemical society.
- **E-Journals - licensed or open access**: E-journal is one which is available in electronic form and can be accessed using computer and communication technologies. It is published and distributed in electronic media. Publisher/ Aggregator is charging some fee to access the resources which are called licensed or paid resources. Some publishers are providing free access to a few of their journals and many organizations are making open access to their products. The e-journals are classified into three types, 1.Online journals, 2.CD-ROM Journals and 3.Networked journals.
- **E-books: Licensed or open access**: E-books are electronic version of books delivered to readers in digital formats. They can be read on all types of computers including handheld devices designed specifically for reading e-books.
- **Reference databases**: Publishers are providing with various reference sources through their websites and databases, such as dictionaries, yearbooks, encyclopaedias etc.
- **Numeric and statistical databases:** It provides historical, financial, statistical and marketing information.
- **Multimedia products:** Multimedia products are finding profound use in education and training, classroom lecturers, operation of a machine, a particular experimentation in a laboratory, a surgery in an operation theatre etc. For using these products a well-configured computer system is needed. The computer should have CD-drive, speakers, adequate memory and so on.

5. The Informal Electronic Resources

- **Blogs**: Blog is an online dairy where one can post information (not only text but also audio, photographs and videos) on a regular basis. Blog is defined as referring content management (or distribution) tool/system which helps to broadcast useful information to end-user in order to promote and create awareness in electronic environment.
- **E-Mail**: User sends an e-mail to library with a reference query whatever information he/she feels is necessary. The library may reply by e-mail, phone, fax, letter, etc. Now days the web pages of libraries are coming with "Ask a librarian" option.

6. Advantages of E-Resources

Most electronic resources come equipped with powerful search-and-retrieval tools that allow users to perform literature searches more efficiently and effectively than was previously possible. Electronic resources are available to users and they can access them for 24 hours a day through the web. Navigate directly from indexing databases to the full text of an article and even follow further links from there. The user can re-specify his /her needs dynamically. The information is obtained based on the need so becomes "just in time "rather than "just in case". Electronic information can therefore provide a number of advantages over traditional print based sources.

The majority of the students have a computer or a laptop at home and access the Internet from home. Students who reside off-campus and who have a cable Internet connection also access the Internet primarily from home. Off-campus students whose Internet connection is through a phone line typically utilize school facilities to do most of their online work. It essentially pervades every aspect of their lives.

The Internet is used by these students for many things, such as to look up information on events, get directions or telephone numbers, get information on products, and to shop. Nearly anything that needs to be done can be done on the Internet. This holds true for college related information as well.

7. Disadvantages of E-Resources

The frequently mentioned disadvantages students reported of electronic resources revolve around the issue of quantity vs. quality of information that is obtained. While many of the mentioned disadvantages also occur with traditional print resources, these students believed that the problems were more pronounced with electronic resources due to the abundance of information they are now dealing with. Electronic resources present the user with vast amounts of information.

It is easy to get distracted or lost on a tangent. A disadvantage is that it is easy to be undisciplined about Internet usage. It is easy to get distracted by all the links and irrelevant information. You end up spending a long time off on tangents. It is hard to determine when to quit searching for information in order to start writing. It provides a lot of time for procrastination. There is too much information. All of the doctoral graduate students mentioned the concern of the Internet connection going down. They so heavily rely on the Internet, that when they lose their connection, they feel at a loss.

8. Growth of Junior College Libraries

Origin of the Department Two year Intermediate Course (+2 level education) hither to Pre-university course is referred to as Junior College was introduced in the year 1969-70. With the introduction of +2 level of Education, there was an enormous increase in the workload in the office of the Director of Public Instruction, and attention to three levels of Education, viz., Secondary, Intermediate and College education could not be given. So the Directorate of Higher Education, which included +2 level of Education, was carved out from the Directorate of Public Instruction, Vide G.O.Ms.No.788, Education Department, dated.30.6.1975. Later, at the time of introduction of UGC Scales of Pay, 1986 to the staff of Composite Degree Colleges, certain problems cropped up. It was decided that the Composite Colleges should be bifurcated into Degree and Junior Colleges.

The main objectives of the Department are to promote Intermediate Education in rural, backward and tribal areas for introducing need based vocational courses as a replacement of conventional courses in a phased manner to create self employment.

9. Aims of the Department are:

1. Administration and control over Government Junior Colleges.
2. Regulating and controlling certain aspects of private aided colleges, vis-a-vis release of grants and utilisation, auditing of accounts and exercising powers envisaged in A.P. Education Act, 1982.
3. Implementation of the scheme of Vocational Education at +2 level, including designing of courses, preparation of reading material and conducting inspections so that by the year 2020, 70% of +2 students are brought under vocational stream.

All the Colleges have Libraries with the in charge of one teacher. There is no separate librarian post sanctioned by the Government in many colleges. In Junior colleges, few librarians were providing library services in traditional methods. There is a need to introduce Electronic information resources in all Junior Colleges libraries for providing better information services to the students as well as teachers.

10. Salient Features of Junior College Libraries

Responsible for the maintenance of Library in the College is;

- Arrange the books in the Library in a scientific manner
- Purchase journals, magazines and newspapers and maintain a reading room for the students
- Maintain the stock register of books and journals supplied to the Library
- Watch the issue of books and journals to various departments and staff members and watch their return
- Prepare the list of books and journals required for the library under the guidance of Library Committee
- Consult various departments and obtain list of books and place an order
- Keep open the Library for the students during the hours fixed by the Principal and issue books to the students
- Behave well with the students and the faculty members
- Responsible for the cleanliness and discipline in the Library.

11. Conclusion

Junior Colleges are special Institutions with an aim to provide the needs of the talented students. The emphasis is laid mainly on providing quality education suited to the present day educational scenario by creating conducive educational atmosphere, help and strives for all round development of the students.

For satisfying the above aims and objectives the junior college libraries should use electronic information resources enormously. As a result of this revolution, the mode of access to the information has shifted from printed paper to electronic format. It should improve in speed, efficiency and effectiveness for providing right information at right time to the right reader.

Currently it is not known how much further the technology age will progress and change in the future. It is remarkable how far in such a short time academic research has identified and looked for solutions to the array of impediments that the information age has had on the success of the community college student. Both librarians and faculty have been resilient and ardent in making changes as technology affects the way that the community college student seeks information. By identifying such problems, we as professionals can use our knowledge and creativeness to find paths around these obstacles. It is a librarian's duty to lead students to a better comprehension of the services and resources that we offer.

References

Michaela Waldman (2003). "Freshmen's use of library electronic resources and self-efficacy" Information Research, Vol. 8 No. 2, January 2003.

Ramaiah L S, Koteswara Rao M, Veeranjaneyulu K and Uma Devi Y, ed. (2011) Library and Information services in the Digital Era, B S Publications. P.P 85-89.

www.ifla.org/files/assets/acguisition-collection-developmet/publications/electronic-resources-guide-2012.pdf.

www.oclc.org/support/services/worldcat/documentation/cataloging/electronic resourcesen.htm?urlm.

www.ingramcontent.com/pod-product-compliance
Ingram Content Group UK Ltd.
Pitfield, Milton Keynes, MK11 3LW, UK
UKHW021532300726
14060UKWH00011B/374

9 789387 057340